CONCISE CHEMISTRY OF THE ELEME...

"Talking of education, people have now a-days" (said he) "got a strange opinion that every thing should be taught by lectures. Now, I cannot see that lectures can do so much good as reading the books from which the lectures are taken. I know nothing that can be best taught by lectures, except where experiments are to be shewn. You may teach chymestry by lectures — You might teach making of shoes by lectures!"

James Boswell: Life of Samuel Johnson, 1766

ABOUT THE AUTHORS

SŁAWOMIR SIEKIERSKI

Sławomir Siekierski gained his Ph.D. in inorganic chemistry from the University of Warsaw in 1955. He became an Associate Professor in 1966, a Full Professor in 1975. He is now at the Institute of Nuclear Chemistry and Technology, where for many years he has headed the Department of Radiochemistry. He has lectured extensively on such subjects as inorganic chemistry, separation methods, and the chemical properties of heavy elements.

His research has focussed mainly on liquid-liquid extraction of metal ions and on the properties of f electron elements. His main achievement in the field of extraction was developing the technique of extraction chromatography and applying it to the separation of radionuclides. His work also contributed substantially to understanding the interactions of metal complexes with organic solvents and with water. He and his collaborators have extensively studied and elucidated the effect of f electron configurations of lanthanide and actinide elements on the stability of their complexes. He is currently working on periodicity in the properties of the elements and on relativistic effects in heavy elements.

JOHN BURGESS

John Burgess's interest in chemistry was kindled and nurtured by three knowledgeable and enthusiastic teachers at Queen Elizabeth's School, Barnet. After an interlude in the Royal Artillery he read Natural Sciences at Sidney Sussex College, Cambridge, following that with Ph.D. research into inorganic solution kinetics with Dr. Reg Prince. Another interlude, at Fisons Fertilizers' Process Development Unit at Levington in Suffolk, was followed by an I.C.I. Fellowship in association with Professor Martyn Symons at the University of Leicester. He has been at Leicester ever since, currently as Emeritus Reader in Inorganic Chemistry. He has lectured at all levels, on such subjects as inorganic solution chemistry, kinetics and mechanisms, analytical and industrial chemistry, oceans and the atmosphere, bioinorganic chemistry, and spectroscopy for biological chemists.

His research has centred on inorganic reaction mechanisms, especially on the use of high pressure techniques, and on solvation of inorganic complexes, particularly those of relevance to human metabolism and to diagnosis and therapy. His *Metal Ions in Solution* (1978) and the shorter *Ions in Solution: Basic Principles of Chemical Interactions* (1988; 2nd edn, 1999) were both published by Ellis Horwood. He has also co-authored *Inorganic Reaction Mechanisms*, with the late Martin Tobe (Addison-Wesley-Longman, 1999), *Template Synthesis of Macrocyclic Compounds*, with Nicolai Gerbeleu and Vladimir Arion (Wiley-VCH, 1999), and *The Colour of Metal Compounds*, with Adam Bartecki (Gordon & Breach, 2000).

Concise Chemistry of the Elements

Slawomir Siekierski
Professor of Chemistry and Radio Chemistry
Institute of Chemistry and Nuclear Technology
Warsaw, Poland

and

John Burgess
Fellow in Inorganic Chemistry
University of Leicester

Horwood Publishing
Chichester

Published in 2002 by
HORWOOD PUBLISHING LIMITED
Coll House, Westergate, Chichester, West Sussex, PO20 6QL England

British Library Cataloguing in Publication Data
A catalogue record of this book is available from the British Library

ISBN 1-898563-71-3

Contents

Periodic Table   viii

Preface   ix

Introduction   1

Part 1 General properties

1 **Many-electron atoms**   5
   1.1 The hydrogen atom and probability density   5
   1.2 Orbital radii   6
   1.3 Many-electron atoms   9
   1.4 Results of the Hartree method   11

2 **Shell filling in many-electron atoms**   15
   2.1 General rules for filling shells in atoms   15
   2.2 Evolution of orbital energy   17
   2.3 Occupation of orbitals and their energy   19

3 **Radii and their changes in the Main-Group Elements**   21
   3.1 Screening from the nuclear charge   21
   3.2 Changes of atomic radii in s and p block elements   22
   3.3 Types of experimental radii   26

4 **Orbital energies and related properties**   33
   4.1 Orbital energies   33
   4.2 Ionization energies   34
   4.3 Electron affinity   35
   4.4 Electronegativity   37
   4.5 Hardness and softness   38
   4.6 Relativistic effects and properties of elements   40

5 **Oxidation states and their stability**   43
   5.1 Valence, hypervalence and oxidation state   43
   5.2 Oxidation states of s and p block elements   45
   5.3 Changes in stability of the maximum oxidation state down the p block Groups   50
   5.4 Dimeric molecules and dimeric cations   53
   5.5 Stereochemical properties of a lone electron pair   54

6 **Catenation and formation of condensed phases**   55
   6.1 Catenation   55
   6.2 Formation of clusters   59
   6.3 Formation of condensed phases   60
   6.4 Conditions for formation of a metallic phase   61
   6.5 The character of the condensed phase and the position of the element in the Periodic Table   64
   6.6 Allotropy   66

Part 2 Chemical properties

7 Hydrogen and the alkali metals   67
   7.1 Properties of hydrogen   68
   7.2 The alkali metals   70

8 Group 2. The alkaline-earth metals   77
   8.1 General properties   77
   8.2 Changes of properties down the Group   78
   8.3 Beryllium   80
   8.4 Structural chemistry   81
   8.5 Coordination and solution chemistry   82

9 Groups 13 and 3   85
   9.1 The Group 13 elements   85
   9.2 Properties of boron   86
   9.3 Properties of Al, Ga, In and Tl   89
   9.4 Group 3: Sc, Y and La   94

10 Group 14   97
   10.1 Properties of carbon   97
   10.2 Properties of Si, Ge, Sn and Pb   100
   10.3 Comparison with Group 4 elements   104

11 Group 15   107
   11.1 Properties of nitrogen   107
   11.2 Properties of P, As, Sb and Bi   110
   11.3 Comparison with Group 5 elements   113

12 Group 16   115
   12.1 Properties of oxygen   115
   12.2 Properties of S, Se, Te and Po   117
   12.3 Comparison with Group 6 elements   119

13 Group 17. The halogens   121
   13.1 Properties of fluorine   121
   13.2 Properties of Cl, Br, I and At   122
   13.3 Structures of ionic halides   126

14 Group 18 (0). The noble gases   127
   14.1 Introduction   128
   14.2 Formation of compounds   128

15 Transition elements   131
   15.1 General characteristics   131
   15.2 The metallic phase   132
   15.3 Oxidation states   134
   15.4 Binary compounds, salts, hydrates and aqua-ions 138
   15.5 Formation of complexes   140
   15.6 Differences between the series   148

16 Group 11. The coinage metals   155
   16.1 General properties   155
   16.2 Changes in properties down the Group   159
   16.3 Unique properties of gold   160

17 Group 12   163
   17.1 General properties   163
   17.2 Changes in properties down the Group   167

18 Lanthanides and actinides 169
  18.1 General characteristics 169
  18.2 The metallic phase 170
  18.3 The sequence of orbital energies and radii 173
  18.4 Lanthanide and actinide contractions 175
  18.5 Oxidation states of lanthanides and actinides 178
  18.6 Binary compounds and salts 180
  18.7 Aqua-cations 183
  18.8 Complex formation and separation 184

19 The transactinide elements 187

20 The structure of the Periodic System 191

Index 196

s-BLOCK d-BLOCK p-BLOCK f-BLOCK

1	2	3	4	5	6	7	8	9	10	11	12	13	14	15	16	17	18
H $1s^1$																	He $1s^2$
Li $2s^1$	Be $2s^2$											B $2s^2 2p^1$	C $2s^2 2p^2$	N $2s^2 2p^3$	O $2s^2 2p^4$	F $2s^2 2p^5$	Ne $2s^2 2p^6$
Na $3s^1$	Mg $3s^2$											Al $3s^2 3p^1$	Si $3s^2 3p^2$	P $3s^2 3p^3$	S $3s^2 3p^4$	Cl $3s^2 3p^5$	Ar $3s^2 3p^6$
K $4s^1$	Ca $4s^2$	Sc $3d^1 4s^2$	Ti $3d^2 4s^2$	V $3d^3 4s^2$	Cr $3d^5 4s^1$	Mn $3d^5 4s^2$	Fe $3d^6 4s^2$	Co $3d^7 4s^2$	Ni $3d^8 4s^2$	Cu $3d^{10} 4s^1$	Zn $3d^{10} 4s^2$	Ga $4s^2 4p^1$	Ge $4s^2 4p^2$	As $4s^2 4p^3$	Se $4s^2 4p^4$	Br $4s^2 4p^5$	Kr $4s^2 4p^6$
Rb $5s^1$	Sr $5s^2$	Y $4d^1 5s^2$	Zr $4d^2 5s^2$	Nb $4d^4 5s^1$	Mo $4d^5 5s^1$	Tc $4d^5 5s^2$	Ru $4d^7 5s^1$	Rh $4d^8 5s^1$	Pd $4d^{10} 5s^0$	Ag $4d^{10} 5s^1$	Cd $4d^{10} 5s^2$	In $5s^2 5p^1$	Sn $5s^2 5p^2$	Sb $5s^2 5p^3$	Te $5s^2 5p^4$	I $5s^2 5p^5$	Xe $5s^2 5p^6$
Cs $6s^1$	Ba $6s^2$	La–Lu	Hf $5d^2 6s^2$	Ta $5d^3 6s^2$	W $5d^4 6s^2$	Re $5d^5 6s^2$	Os $5d^6 6s^2$	Ir $5d^7 6s^2$	Pt $5d^9 6s^1$	Au $5d^{10} 6s^1$	Hg $5d^{10} 6s^2$	Tl $6s^2 6p^1$	Pb $6s^2 6p^2$	Bi $6s^2 6p^3$	Po $6s^2 6p^4$	At $6s^2 6p^5$	Rn $6s^2 6p^6$
Fr $7s^1$	Ra $7s^2$	Ac–Lr	Rf $6d^2 7s^2$	Db $6d^3 7s^2$	Sg $6d^4 7s^2$	Bh $6d^5 7s^2$	Hs $6d^6 7s^2$	Mt $6d^7 7s^2$	110 $6d^8 7s^2$	111 $6d^9 7s^2$	112 $6d^{10} 7s^2$						

f-BLOCK

La $5d^1 6s^2$	Ce $4f^1 5d^1 6s^2$	Pr $4f^3 6s^2$	Nd $4f^4 6s^2$	Pm $4f^5 6s^2$	Sm $4f^6 6s^2$	Eu $4f^7 6s^2$	Gd $4f^7 5d^1 6s^2$	Tb $4f^9 6s^2$	Dy $4f^{10} 6s^2$	Ho $4f^{11} 6s^2$	Er $4f^{12} 6s^2$	Tm $4f^{13} 6s^2$	Yb $4f^{14} 6s^2$	Lu $4f^{14} 5d^1 6s^2$
Ac $6d^1 7s^2$	Th $6d^2 7s^2$	Pa $5f^2 6d^1 7s^2$	U $5f^3 6d^1 7s^2$	Np $5f^4 6d^1 7s^2$	Pu $5f^6 7s^2$	Am $5f^7 7s^2$	Cm $5f^7 6d^1 7s^2$	Bk $5f^9 7s^2$	Cf $5f^{10} 7s^2$	Es $5f^{11} 7s^2$	Fm $5f^{12} 7s^2$	Md $5f^{13} 7s^2$	No $5f^{14} 7s^2$	Lr $5f^{14} 6d^1 7s^2$

Preface

This book shows how fundamental properties of atoms affect chemical and some physical properties of elements and, in particular, how and why these properties change as one traverses the systematic grouping of the elements in the Periodic Table. The fundamental properties of atoms, selected for this purpose, are radii and energies of outer orbitals, ionization potentials and electron affinities. Attention is focused upon unique properties of the first s, p and d shells and on the effect of filled $3d$ and $4f$ shells on the properties of p and d block elements, respectively. Special attention is also paid to the role of relativistic effects in the chemistry of heavy elements. The book is divided into two parts. In Part 1 changes in fundamental properties of atoms across Groups and Periods are reviewed, whereas Part 2 provides a concise description of chemical properties of elements.

In Chapter 1 orbital radii and energies are introduced and their properties in many-electron atoms are briefly discussed. Chapter 2 is devoted to the evolution of orbital energies, and explains how competition between the n and l quantum numbers affects filling of atomic orbitals. Chapter 3 shows how radii of atoms change in the Periodic Table and how covalent, metallic and ionic radii depend on the radii of outer shells. In Chapter 4 changes of orbital energies, ionization potentials, electron affinity and electronegativity in the Periodic Table are discussed. It also contains a simple introduction to relativistic effects in atoms and their role in determining chemical properties of elements. Chapter 5 is a review of oxidation states of the elements, and answers the question why oxidation numbers change in a different way for p and d block elements. The last (sixth) Chapter of Part 1 provides information on how and why the character of the condensed phase changes from element to element. This Chapter also deals with the ability of atoms to catenate and form clusters, and shows the correlation between catenation and formation of condensed phases.

Chapters 7 to 14 of Part 2 provide presentations of the basic chemical properties of the Main Group elements. Attention is focused upon correlations between chemical properties of elements and fundamental properties of atoms. Evidence for non-uniform changes of properties down the Main Groups is presented, and their origin discussed. Chapters 15 to 17 are devoted to d electron and Groups 11 and 12 elements. Again, special attention is paid to differences between the first element in a d block Group and its heavier congeners. The role of the filled $4f$ shell and of relativistic effects in differentiating chemical properties of d elements is discussed. In Chapter 18 the chemistry of lanthanides and actinides is covered briefly from the viewpoint of the unique properties of f orbitals with respect to energy and radial extent. Chapter 19 is devoted to the known and predicted properties of transactinide elements. Finally Chapter 20 recapitulates changes of chemical properties of elements within blocks, with special emphasis on the primary and secondary structure of the Periodic Table.

The approach presented in the book is qualitative and requires from the reader only rudimentary knowledge of quantum chemistry. The book can be used by both undergraduate and postgraduate students as a guide and introduction to comprehensive textbooks on inorganic chemistry.

Data sources

Values for orbital energies and radii have been taken from J. P. Desclaux, *Atomic Data and Nuclear Data Tables* **12**, 311 (1973). Orbital values of lanthanide and actinide ions were kindly calculated by Dr. W. Jaskólski of the Department of Physics, Nicolaus Copernicus University, Toruń, Poland. Ionization energies, electron affinities, electronegativities, atomization energies, melting and boiling points of elements, and reduction potentials are from J. Emsley, *The Elements*, Oxford University Press (1991). The values quoted for ionic radii are those tabulated in R. D. Shannon, *Acta Cryst.* **A32**, 751 (1976). Hydration enthalpies of ions are from Y. Marcus, *Ion Solvation*, John Wiley & Sons (1985) or from J. Burgess, *Ions in Solution*, 2nd edn., Horwood Publishing (1999).

Acknowledgements

The authors would like to thank first of all Professor Jerzy Narbutt for persuading them to undertake the preparation of this revised, updated, and enlarged English edition of SS's *Chemia pierwiastków*. They would also like to express their gratitude for the interest, encouragement, and support provided by Ellis Horwood and by Anna Raiter-Rosińska of Polish Scientific Publishers PWN. SS wishes to express his gratitude to colleagues in the Radiochemistry Department for their support, interest, and help in updating the book for this new edition. JB wishes to express his appreciation to Professor Samotus and her group in Kraków and to Profesor Bartecki in Wrocław for their most generous hospitality during his visits to Poland in connection with book writing and collaborative research. JB would also like to record his gratitude to Dr and Mrs Grzedzielscy of Ipswich for introducing him many years ago to Polish hospitality and customs, thereby encouraging him to visit Poland when the opportunity arose.

John Burgess   Sławomir Siekierski
Department of Chemistry   Department of Radiochemistry
University of Leicester   Institute of Nuclear Chemistry
Leicester LE1 7RH   and Technology
U.K.   ul Dorodna 16
                                     30-195 Warsaw
                                        Poland

Introduction

The concept of elements originates from the ancient Greek philosophy of nature. According to Empedocles (5th century B.C.) all matter was built from four primordial "elements" fire, air, water and earth; these elements were united and parted by two "active forces", love and strife. Up to the 17th century only 13 elements in the modern meaning of this word were known, where by known we mean that they were used in a relatively pure state. Starting from the second half of the 18th century an avalanche of discoveries of elements started, which has lasted until now. At present (at the start of the year 2002) we know for sure that 112 elements exist. In this connection it is worth to ponder briefly the question what are now the criteria of discovery of a new element. It is commonly agreed that discovery of a new element requires establishing its atomic number, Z, either by chemical methods or by means of atomic physics (X-ray spectroscopy and other techniques). The isotope's mass number may in general remain unknown. Chemical identification is an ideal proof that a new element has been synthesized. In the case of short-lived nuclides produced in amounts of a few atoms per experiment two requirements must be met. First, the chemical method must be fast and capable of giving the same results for one-atom-at-a-time as would be obtained in experiments with macroscopic amounts. This criterion is met by e.g. thermal gas chromatography, extraction and ion exchange chromatography. This is because in these methods one atom is involved in many identical operations, which is equivalent to a large number of atoms participating in only one operation. The second requirement is that the presence of the new element in the appropriate chemical fraction is unequivocally established by observing its decay mode. Chemical procedures have been used in identification of heavy actinides and transactinides up to element 108 (hassium). On the other hand only the application of nuclear physics techniques proved the discovery of elements with atomic number $Z > 108$. These techniques consisted in measuring energies of α particles and half-lives of the new isotope and of the daughters in a chain of quickly following α decays, which ends with a known nuclide. For instance element 112 was discovered on the basis of the following chain:

$$^{277}112 \xrightarrow{\alpha} {}^{273}110 \xrightarrow{\alpha} {}^{269}108 \xrightarrow{\alpha} {}^{265}106 \xrightarrow{\alpha} {}^{261}104 \xrightarrow{\alpha} {}^{257}102$$

The nuclides $^{265}106$, and $^{261}104$ and $^{257}102$ were known from previous experiments and are known isotopes of chemically identified elements. The initial nuclide in this chain was obtained by bombarding a target of the 208 lead isotope with neutron-rich zinc nuclides:

$$^{208}_{82}Pb_{126} + {}^{70}_{30}Zn_{40} \longrightarrow {}^{277}112_{165} + n$$

Of the 112 established elements 22 are man-made. The latter include technetium, promethium, and the elements from neptunium to element 112. However, it should be noted that trace amounts of ^{99}Tc, ^{147}Pm, ^{239}Np and ^{239}Pu are found in uranium ores and the presence of primordial ^{244}Pu ($\tau_{1/2} = 8 \times 10^7$ years) in the earth's crust has been reported. The man-made elements have been produced in several types of nuclear reactions:

 – Fission of uranium, which yielded technetium and promethium
 – Neutron capture (n,γ) reactions followed by β^- decay :

$$^AZ + n \rightarrow {}^{A+1}Z + \gamma$$

$$^{A+1}Z \rightarrow {}^{A+1}(Z+1) + \beta^-$$

Elements from Np to Am have been synthesized by such reactions. Still heavier transactinides, which require more neutrons in their nuclei to be stable, have been obtained in (n,γ) reactions at high neutron fluxes. Under such conditions absorption of neutrons becomes faster than the β^- decay and many successive (n,γ) reactions may occur before the nuclides undergo β^- decay. Einsteinium and fermium were first obtained in this way during testing of thermonuclear devices.

 – Bombardment of plutonium and transplutonium targets with helium or with carbon, oxygen, neon and magnesium nuclides. Examples are:

$$^{242}_{96}\text{Cm} + {}^{4}_{2}\text{He} \longrightarrow {}^{245}_{98}\text{Cf} + n$$

$$^{248}_{96}\text{Cm} + {}^{18}_{8}\text{O} \longrightarrow {}^{259}_{102}\text{No} + \alpha + 3n$$

Using this type of reaction elements up to element 108 have been synthesized.

 – Bombardment of lead or bismuth targets with Zn or Ni nuclides. The elements from 107 to 112 were first produced this way. An example is the synthesis of element 112 shown above.

The elements which are found on the surface and in the deeper layers of the earth can be divided into two groups. To the first group belong elements whose nuclei are essentially stable or have very long half-lifes ($\tau_{1/2} > 10^8$ years). To the second group belong those elements that are short-lived but are members of a radioactive decay series. An example is radon (^{222}Rn, $\tau_{1/2} = 3.82$ days) which is the member of the uranium decay series. The stable and the very long-lived elements have been formed in two different stages of the universe's evolution: in the early stage before stars were formed and at the late stage simultaneously with the formation of stars. In the first stage, which lasted from about 100 seconds to about 10^5 years after the big bang, the temperature was initially about 10^9 K, which made possible formation of deuterons and afterwards ^{4}He nuclei in the following reactions:

$$p + n \rightarrow d + \gamma$$

$$d + d \rightarrow {}^3\text{He} + n$$

$$^3\text{He} + n \rightarrow {}^4\text{He} + \gamma$$

At the end of this stage the temperature dropped to about $3 \times 10^3 K$, which brought nuclear reactions to a standstill. About 75% of matter then consisted of protons and about 25% of ^4He. In the second stage stars began to form and their gravitational contraction increased the temperature above $10^7 K$, which made possible fusion of protons into ^4He by the net reaction

$$4p \rightarrow {}^4He + 2e^+ + 2\nu_e + \Delta E$$

where ν_e is the electron neutrino. This process is called hydrogen burning. With increase of the temperature to about $10^8 K$ the accumulated ^4He nuclei were able to undergo further thermonuclear reactions, called helium burning :

$$^4He + {}^4He \rightarrow {}^8Be + \gamma$$
$$^8Be + {}^4He \rightarrow {}^{12}C + \gamma$$

The two processes, i.e. hydrogen and helium burning, lasted from about 10^9 to about 5×10^9 years. At increasing concentration of ^{12}C in the core of the stars synthesis of heavier nuclei began, e.g.

$$^{12}C + {}^4He \rightarrow {}^{16}O + \gamma$$

With further increase of temperature fusion of two ^{12}C nuclei (carbon burning) or two ^{16}O nuclei (oxygen burning) becomes possible, leading to formation of e.g. ^{24}Mg and ^{28}Si nuclei, respectively (in the latter case with emission of an α particle). Synthesis of still heavier nuclei up to $A = 56$ began at temperatures $\geq 10^9$ K by e.g. fusion of ^{28}Si nuclei. With these reactions the thermonuclear syntheses comes to an end. The reason is that binding energy per one nucleon attains its maximum at around $A = 56$ so that heavier nuclei become unstable and cannot be formed at such high temperatures. Therefore, other processes had to enter the scene. At very high temperatures excited nuclei emit γ photons which liberate neutrons from the nuclei by (γ, n) reactions. At low neutron fluxes, neutron absorption is slower than β^- decay and unstable nuclides formed by (n, γ) reactions have enough time to change by β^- decay into stable nuclides. Formation of elements from iron to bismuth can be explained by this process, called the slow process. At very high neutron flux densities, as in supernova explosions, absorption of neutrons becomes faster than β^- decay. Formation of ^{232}Th and ^{238}U can only be explained by the rapid process, the same as that used by man to produce Es and Fm.

Of the 112 known elements, 97 meet the narrow chemical criterion for discovery of an element, i.e. occur in nature or have been produced in nuclear reactions in weighable amounts. These are elements from hydrogen to fermium minus francium (^{223}Fr, $\tau_{1/2} = 22$ minutes), astatine (^{210}At, $\tau_{1/2} = 8.1$ hours) and radon (^{222}Rn, $\tau_{1/2} = 3.8$ days). Under normal conditions 10 of the 97 elements are gases (H, He, Ne, Ar, Kr, Xe, N, O, F, Cl), 2 are liquids (Br and Hg, and at above 30 °C also Ga and Cs) and 85 form solid phases. The elements, even the metallic elements, differ widely in their melting and boiling points. For instance, the melting point of Hg is −38.9 °C, while that of tungsten is about 3420 °C. Of the 87 elements which under normal conditions form condensed phases 77 show metallic properties, 4 are semiconductors (Si, Ge, Se, Te) and 6 are non-metals (B, C, P, S, Br, I). Hydrogen and the gaseous p block elements (N, O, F, Cl and noble gases) are also commonly classified as non-metals. If we take into account that At, Fr, and the elements from 101 to 112 would form metallic phases if not for their nuclear properties, then the share of metals in the total number of elements becomes even greater. In this connection it is worth noting that except for hydrogen all non-metals and semi-conductors are p block elements. The main reason is that p elements are end-elements in each row, where the effective nuclear charge acting on valence electrons is high. High effective nuclear charge results in high ionization potentials which, in turn, renders difficult formation of a metallic phase.

Elements differ very much in some of their physical properties e.g. they span a tremendous range of electrical resistivity, as shown in Table I.1. From the chemist's point of view it is important that elements differ as much in their chemical as in their physical properties. For instance, the complete inertness of helium should be compared with the extreme chemical reactivity of fluorine. Metallic elements also differ widely in chemical reactivity, from the inertness of gold, which dissolves only in aqua regia, to the high reactivity of the alkali metals, which must be handled in an inert atmosphere. It is also interesting to notice abrupt changes in properties between neighbouring elements in a row, i.e. between elements which differ by only one electron in the same valence subshell. A most striking example is the difference between carbon and nitrogen with respect to the stable form of the element under normal conditions, which for carbon is a solid of very high melting point and for nitrogen a diatomic gas. Explaining how and why properties of orbitals in atoms, and their filling by electrons, determine chemical properties of elements is the main aim of what is conventionally called "the chemistry of elements".

Table I.1 – Electrical resistivity of some elements, in Ω m.

Ag	Bi	Te	Ge	B	P^a	S
1.6×10^{-8}	1.1×10^{-6}	4.4×10^{-3}	4.6×10^{-1}	1.8×10^4	10^9	2×10^{15}

^{a} White phosphorus.

PART 1 : *General properties*

1

Many-electron atoms

1.1 THE HYDROGEN ATOM AND PROBABILITY DENSITY.

From the solution of the Schrödinger equation for the hydrogen atom the following expression for the energy, in eV, is obtained:

$$E_n = -\frac{13.6}{n^2} \qquad n = 1, 2, 3 \ldots \tag{1.1}$$

and for the electron wave function or orbital

$$\psi = R_{nl}(r)Y_{lm}(\theta,\varphi) . \tag{1.2}$$

In these expressions n denotes the principal quantum number, l the orbital momentum quantum number and m the magnetic quantum number. $R_{nl}(r)$ is the radial part of the wave function, which determines the variation of the orbital with distance from the nucleus. The angular wave function $Y_{lm}(\theta,\varphi)$ expresses the orbital angular shape. From eq. 1.1 the energy of an electron in the hydrogen atom depends on the principal quantum number only. However, in the spectrum of the hydrogen atom the so-called fine splitting of energy levels is observed. This depends on the quantum number j but is about 10^4 times smaller than the distance between the levels and is a relativistic effect.

In order to understand the properties of orbitals in the hydrogen atom and in many-electron atoms it is of primary importance to know the probability of finding an electron in the vicinity of the nucleus. It follows from the basic interpretation of the wave function, ψ, that the probability $P_{nl}(r, \theta, \varphi)\mathrm{d}V$ of finding an electron with quantum numbers nl in a space element of small volume $\mathrm{d}V$ around a point characterized by spherical coordinates r, θ, φ is given by

$$P_{nl}(r, \theta, \varphi)\mathrm{d}V = |\psi_{ni}(r, \theta, \varphi)|^2 \mathrm{d}V \tag{1.3}$$

However, of more importance is the probability of finding an electron at a certain distance, r, from the nucleus irrespective of direction, rather than at a given point. The probability of finding an electron in a spherical shell of radius r and thickness $\mathrm{d}r$ is denoted by $P_{nl}(r)\mathrm{d}r$, where $P_{nl}(r)$ is the radial probability density, which depends only on r. Fig. 1.1 shows the difference between the probability of finding an electron in a volume dV, $P_{nl}(r, \theta, \varphi)dV$, and in a spherical shell of thickness dr, $P_{nl}(r)dr$.

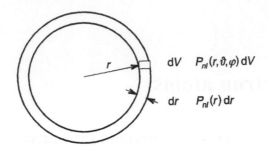

Fig. 1.1 Probability of finding an electron in a volume dV and in a spherical shell of thickness dr.

It can easily be shown that

$$P_{nl}(r) = R_{nl}^{2}(r)\, r^2 . \qquad (1.4)$$

In the vicinity of the nucleus, i.e. for $r \to 0$,

$$R_{nl}(r) \propto (r/a_0)^l \qquad (1.5)$$

where l is the orbital momentum quantum number and a_0 denotes the radius of the first Bohr orbit, equal to 52.9 pm. Hence

$$P_{nl}(r) \propto (r/a_0)^{2l} r^2 \qquad (1.6)$$

Since for a small distance from the nucleus $r \ll a_0$ and $r/a_0 \ll 1$, the radial probability density decreases with increasing quantum number l. This result is valid for hydrogen atoms as well as for many-electron atoms. The reason is that for an electron which is in the vicinity of the nucleus interactions with the remaining electrons cancel (see Fig. 1.2).

1.2 ORBITAL RADII

With the aid of the function $P_{nl}(r)$ one can define two different radii (Fig. 1.3) for a given orbital:

- $r_{\max,nl}$, which is the distance from the nucleus corresponding to the maximum radial probability density, and
- $\langle r_{nl} \rangle$, which is the expected (average) value of r:

$$\langle r_{nl} \rangle = \int_{0}^{\infty} r\, P_{nl}(r)\, dr. \qquad (1.7)$$

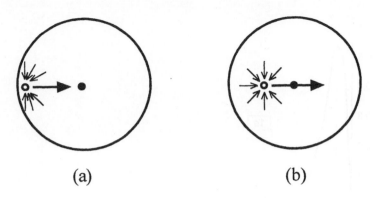

Fig. 1.2 Forces acting on an electron in a many-electron atom: (a) at the surface of the atom, (b) in the vicinity of the nucleus.

In all cases $r_{\max,nl} < \langle r_{nl} \rangle$. Each of the two radii may serve as a measure of the radial extent of an orbital and of the subshell to which this orbital belongs. The radii $\langle r_{nl} \rangle$ and $r_{\max,nl}$ depend on the quantum numbers n and l; radii $\langle r_{nl} \rangle$ (and also $r_{\max,nl}$) conform to the following rules:

– In the case of the hydrogen atom and many-electron atoms, for a given quantum number l the orbital radius increases with increasing n, for instance for $l = 0$ (Fig. 1.3):

$$\langle r_{1s} \rangle < \langle r_{2s} \rangle < \langle r_{3s} \rangle$$

– For the hydrogen atom and for a given quantum number n the orbital radius decreases with increasing l, e.g. for $n = 3$ (Fig. 1.3) we have:

$$\langle r_{3s} \rangle > \langle r_{3p} \rangle > \langle r_{3d} \rangle$$

– In many-electron atoms, for constant quantum number n, the orbital radius increases with increasing l in the case of outer orbitals and changes only very slightly in the case of inner orbitals (see Section 1.4).

We know that in the hydrogen atom and in hydrogen-like atoms (atoms with only one electron) the electron energy depends solely on n and is independent of l, while the orbital radius depends on both quantum numbers n and l (Fig. 1.3). Thus, the same orbital energy corresponds to different average distances from the nucleus, which seems to be at variance with intuition. The independence of the total electron energy in hydrogen-like atoms from the shape of the orbital (from the quantum number l) results from the $1/r$ character of the potential in the Coulomb field. In a field with such characteristics there is a subtle equilibrium between potential and kinetic energy which makes the total energy constant, in spite of variation of the orbital shape. The same is also a characteristic feature of the gravitational field. Qualitatively this can be explained by comparing the $2s$ with the $2p$ electron in the (excited) hydrogen atom. According to eq. (1.6) the probability of finding an s electron ($l = 0$) close to the nucleus is high. Being near the nucleus the $2s$ electron experiences very strong attraction, which makes its orbital energy more negative than that of the $2p$ electron ($l = 1$), which appears less frequently in the vicinity of

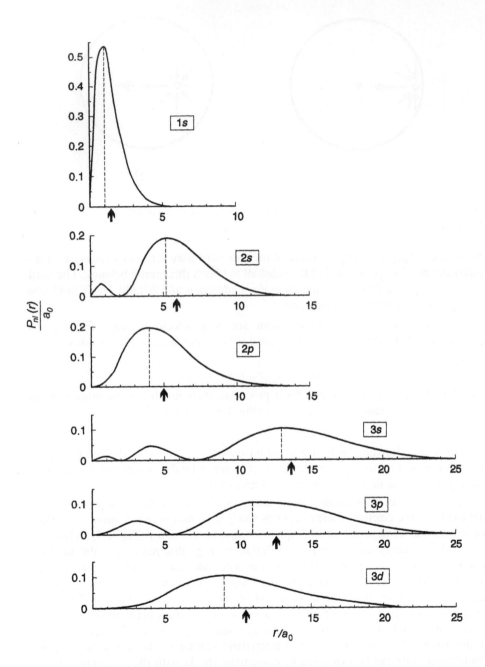

Fig. 1.3 Radial probability density for an electron in the hydrogen atom. The arrow denotes the value of the radius $\langle r_{nl} \rangle$.

the nucleus. However, the $2s$ electron can lose the excess of its binding energy by making distant "excursions" away from the nucleus, where the nuclear attraction is weak. Such excursions make the orbital radius of the $2s$ electron large. Instead the $2p$ electron, which more rarely penetrates the atomic core, can keep its energy equal to that of the $2s$ electron without moving far away from the nucleus. Therefore, the orbital energies of the two electrons can be equal, $\varepsilon_{2s} = \varepsilon_{2p}$, but at the same time the relation $\langle r_{2s} \rangle > \langle r_{2p} \rangle$ must hold.

1.3 MANY-ELECTRON ATOMS

The potential energy of the i-th electron in a many-electron atom is given by

$$V_i(r_i) = -\frac{Ze^2}{4\pi\varepsilon_0 r_i} + \sum_{\substack{i=1 \\ j>i}}^{n} \frac{e^2}{4\pi\varepsilon_0 r_{ij}} \qquad (1.8)$$

where e denotes the elementary charge, Z the atomic number, n the number of electrons, ε_0 the dielectric permeability of vacuum, r_i the distance of the i-th electron from the nucleus and r_{ij} the distance between the i-th and j-th electron. Because of the r_{ij} term the Schrödinger equation cannot be solved analytically. Therefore, in order to determine the orbitals in a many-electron atom we use the so-called one-electron approximation, of which the simplest form is the Hartree method. In this method we assume that each electron in a many-electron atom is moving independently in a spherically symmetrical potential, $V(r)$, which results from superposition of the spherically symmetrical potential of the nucleus and the spherically symmetrical potential of the remaining electrons. The electrons are on the average spherically distributed because of mutual repulsion.

The Hartree method consists of the following steps:

1. The approximate form of the $V(r)$ potential acting on a typical electron is ssumed:

In order to establish the approximate dependence of the potential on r it is assumed that the charge acting on the electron far away from the nucleus is equal to $+e$, while that acting on an electron in the vicinity of the nucleus is equal to $+Ze$. This is because at large distances the positive charge of the nucleus is almost completely compensated by the negative $-(Z-1)e$ charge of the rest of electrons. On the other hand in the vicinity of the nucleus the action of the remaining electrons cancels because of their spherical distribution, and the electron in question is under the full $+Ze$ charge of the nucleus (see Fig. 1.2). Table 1.1 shows the assumed effective nuclear potential acting on the electron as a function of the distance.

2. The Schrödinger equation for the typical electron is solved numerically:

We obtain the set of orbitals $\psi_\alpha(r, \theta, \varphi)$, $\psi_\beta(r, \theta, \varphi)$, ... and orbital energies ε_α, ε_β, ... where α, β ... denote full sets of quantum numbers.

Table 1.1 – Effective nuclear charge and potential acting on the electron as a
function of distance r from the nucleus in a many-electron atom.

Distance	Effective charge	Potential
$r \to 0$	$+ Ze$	$V(r) = - \dfrac{Ze^2}{4\pi\varepsilon_0 r}$
$r \to \infty$	$+ Ze - (Z-1)e = +e$	$V(r) = - \dfrac{e^2}{4\pi\varepsilon_0 r}$
$0 < r < \infty$		Reasonable interpolation*

* Reasonable interpolation takes into account that the rate with which the potential changes
decreases with increasing distance from the nucleus.

3. Orbitals are filled by electrons:
Orbitals are filled in the way which secures minimum total energy of the atom and
is in agreement with the Pauli exclusion principle.

4. The distribution of charge density is determined:
From the orbitals calculated in step 2 we determine the charge distribution of the
$n - 1$ electrons interacting with the typical electron

$$- e \sum_{\substack{j=1 \\ j \neq i}}^{n} |\psi_i(r)|^2 \tag{1.9}$$

5. The potential of the electrostatic field, $V(r)$, is calculated:
From the distribution of the electron charge and from the charge of the nucleus we
determine with the help of the Gauss law the electrostatic field and then its
potential, $V(r)$. The potential calculated in this way differs from that assumed in step
1 and is more realistic.

6. The procedure is repeated starting from the second step

$$1 \rightarrow 2 \rightarrow 3 \rightarrow 4 \rightarrow 5$$

and is continued until the difference between the consecutive values of the atom
energy becomes less than a certain (small) value.

1.4 RESULTS OF THE HARTREE METHOD

1. As in the hydrogen atom the orbital in many-electron atoms, called in this case a Hartree orbital, is a product of the radial and angular functions (see eq. 1.2).

2. Because of the spherical symmetry of the charge distribution the form of the angular function, $Y_{lm}(\theta, \varphi)$, is the same as in the hydrogen atom.

3. Because of repulsion between electrons the form of the radial function, $R_{nl}(r)$, differs from that in the hydrogen atom, except for small distances from the nucleus where it is the same:

$$\text{for } r \to 0, \ R_{nl}(r) \propto (r/a_0)^l \qquad (1.10)$$

Therefore, in the vicinity of the nucleus

$$P_{nl}(r) = R_{nl}^2(r) r^2 \propto (r/a_0)^{2l} r^2 \qquad (1.11)$$

and the probability function shows the same dependence on l and r as in the case of the hydrogen atom.

4. The energy of the Hartree orbital, ε_{nl}, depends on both quantum numbers n and l, not only n as in the case of the hydrogen atom. This is because interaction between electrons removes degeneracy. For a many-electron atom we have:

   – For a given l the orbital energy (which is a negative quantity), increases with increasing n, as in the hydrogen atom

$$\varepsilon_{nl} < \varepsilon_{(n+1)l} < \varepsilon_{(n+2)l} \ \cdots \cdots$$

   – For a given n orbital energy increases with increasing l

$$\varepsilon_{ns} < \varepsilon_{np} < \varepsilon_{nd} \ \cdots \cdots$$

The reason for the latter sequence is the decreasing radial probability density near the nucleus with increasing quantum number l, see eq.(1.11). However, the effect of the quantum number l depends on how big the principal quantum number is. For low values of n, i.e. for innermost orbitals, the effect of l is small. This is because the interelectronic repulsion partially cancels and the electron in an inner shell of a many-electron atom behaves as in a hydrogen-like atom. For higher values of n, i.e. for outer orbitals, the effect of l on orbital energy is much greater, because for an outer electron repulsion with the rest of the electrons does not cancel, see Fig. 1.2. The effect of quantum numbers n and l on the orbital energy and radius in the krypton atom is illustrated by the data in Table 1.2. One can see in this Table that the energy difference between the $2s$ and $2p$ orbitals is smaller than that between the $3s$ and $3p$ orbitals and much smaller than that between the $4s$ and $4p$ orbitals. Table 1.2 also shows that in the krypton atom the $2s$ orbital energy is more negative than $2p$ orbital energy in spite of the greater radius of the former, which means that the two orbitals behave to some extent like orbitals in hydrogen-like atoms. However, in such atoms the $2s$ and $2p$ orbitals are of the same energy. On the other hand for the outer $4s$ and $4p$ orbitals the increase in orbital energy is paralleled by the increase in the orbital radius.

Table 1.2 – Orbital energies and orbital radii (in atomic units) in the krypton atom (E_h = 27.21 eV; a_0 = 52.9 pm).

Orbital	1s	2s	2p	3s	3p	4s	4p
ε_{nl} / E_h	− 529	− 72.0	− 62.9	− 11.2	− 8.41	− 1.19	− 0.52
$\langle r_{nl} \rangle / a_0$	0.04	0.18	0.16	0.53	0.59	1.60	1.94

It is interesting to compare the effect of quantum numbers n and l on the electron energy of lithium and sodium atoms in excited states, i.e. on the energy of the outermost orbitals, Fig. 1.4. One can first see in this Figure that for the lithium atom the orbital energy is always less negative for higher values of the quantum number n, irrespective of the value of l. The reason is the low charge of the lithium nucleus. Because of this, even the s electron which penetrates the atomic core deeply (see eq. 1.11) is not stabilized enough to reverse the trend of increasing orbital energy with increasing n. On the other hand the higher nuclear charge of the sodium atom, together with the much higher probability density near the nucleus of the s than of the d electron reverse the trend, so that for sodium we have $\varepsilon_{4s} < \varepsilon_{3d}$ and $\varepsilon_{5s} < \varepsilon_{4d}$. In both lithium and sodium atoms, for a given n, the energy increases with increasing l. However, this increase is smaller the greater is n. That is because the radii of the outermost orbitals are so large that even the s electron has little chance to come close to the nucleus. Therefore, the energy of the outermost vacant orbitals in many-electron atoms becomes almost independent of l, as in the hydrogen atom.

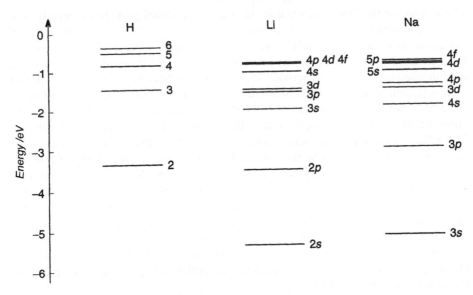

Fig. 1.4 Some energy levels in the H, Li and Na atoms.

5. The total energy of a many-electron atom is not equal to the sum of orbital energies $\Sigma \varepsilon_{nl}$. This is because in calculating the potential acting on the i-th electron we take into account its interaction with all electrons, including the j-th electron. At the same time in calculating the potential acting on the j-th electron we take into account its interaction with the i-th electron. Therefore the sum $\Sigma \varepsilon_{nl}$ contains the interaction for each pair of electrons counted twice.

6. For a given value of l, $\langle r_{nl} \rangle$ increases with n (as in the hydrogen atom), whereas, for a given value of n, $\langle r_{nl} \rangle$ remains almost constant or even decreases with increasing l for inner orbitals (the decrease in the hydrogen atom is very large) and increases for outer orbitals. For instance, in the krypton atom (Table 1.2) we have

$$\langle r_{2s} \rangle > \langle r_{2p} \rangle$$

whereas

$$\langle r_{3s} \rangle \approx \langle r_{3p} \rangle \quad \text{and} \quad \langle r_{4s} \rangle < \langle r_{4p} \rangle .$$

The reason is the same as that in the case of orbital energies. For inner orbitals the decisive role is played by the interaction of the electron with the nuclear charge, hence the sequence of radii resembles that in the hydrogen atom. In contrast, in the case of the outer orbitals the interelectronic repulsion becomes of importance, which results in a reversed order of orbital radii.

In many-electron atoms the orbital energy is a function of the quantum number l for two reasons. The first is the dependence on l of the radial probability density in the vicinity of the nucleus, and the second is the non-Coulomb form of the potential in the intermediate region between the nucleus and the surface of the atom. The potential in the intermediate region is not coulombic, i.e. is not proportional to $1/r$, because the effective nuclear charge which acts on the electron changes with the distance from the nucleus. The dependence of orbital energy on l in many-electron atoms can be qualitatively explained in the following way. As the probability of finding the s electron ($l = 0$) near the nucleus is greater than that of the p electron ($l = 1$), the s electron is more strongly bound. However, in the case of the hydrogen atom this gain in binding energy is compensated by "excursions" of the s electron away from the nucleus, where its binding energy is small. In contrast, such compensation is not possible in many-electron atoms. The reason is that in a many-electron atom the absolute value of the potential energy is very high in the vicinity of the nucleus not only because of the $1/r$ dependence but also because the effective nuclear charge increases from $+e$ to $+Ze$ with decreasing r (Table 1.3). Therefore, no excursions of the s electron, even far away from the nucleus, are able to compensate for the gain in binding energy in the vicinity of the nucleus. This makes the absolute value of the orbital energy in many-electron atoms decrease and the orbital radius increase (slightly for inner orbitals) with increasing quantum number l. The very low lying orbitals with quantum number $n = 2$ provide an exception to the latter trend, see Table 1.2.

Table 1.3 – The dependence of the potential on the distance r from the
nucleus in the hydrogen atom and in a many-electron atom

		distance	
		small	large
Potential	in the hydrogen atom	$-\dfrac{e^2}{4\pi\varepsilon_0 r}$	$-\dfrac{e^2}{4\pi\varepsilon_0 r}$
	in a many-electron atom	$-\dfrac{Ze^2}{4\pi\varepsilon_0 r}$	$-\dfrac{e^2}{4\pi\varepsilon_0 r}$

2

Shell filling in many-electron atoms

A shell is a set of orbitals having the same principal quantum number n; a subshell is a set of orbitals having the same pair of quantum numbers nl. The number of orbitals in a subshell is $2l + 1$, the capacity for electrons $2(2l + 1)$. The electron configuration of an atom gives the number of electrons in the subshells. For instance the configuration of the argon atom is

$$[Ne]3s^2 3p^6$$

Coupling of orbital and spin angular momentum of the electron gives the total angular momentum, characterized by the corresponding quantum number j, $j = l \pm \frac{1}{2}$. Because of interaction between spin and orbital angular momentum the p orbitals split into $p_{1/2}$, $p_{3/2}$, d into $d_{3/2}$, $d_{5/2}$ and f into $f_{5/2}$, $f_{7/2}$ orbitals. The maximum number of electrons in the split l subshell is $2j + 1$. For heavy atoms the j orbitals differ considerably in energies and radii. This is because spin-orbit splitting is a relativistic effect and increases approximately as Z^2. The orbital energy is always more negative and the orbital radius smaller for $j = l - \frac{1}{2}$ than for $j = l + \frac{1}{2}$. For instance for the Bi atom, which has the ground-state electron configuration

$$[Xe]4f^{14}5d^{10}6s^2 6p_{1/2}{}^2 6p_{3/2}{}^1,$$

the energy of the $6p_{1/2}$ orbital is lower than that of the $6p_{3/2}$ orbital by 2.1 eV, and the difference between the respective radii is about 21.5 pm. Because of the presence of one electron in the less stable $p_{3/2}$ orbital, bismuth shows the oxidation number +1 in addition to the common +3 and +5.

2.1 GENERAL RULES FOR FILLING SHELLS IN ATOMS

The ground-state electron configurations of many-electron atoms are experimental characteristics, but can also be predicted from orbital energies together with the Pauli exclusion principle. According to the so-called building-up (Aufbau) principle, electrons occupy orbitals doubly in the order of increasing orbital energies. It follows from calculations by the Hartree method that the sequence of orbital energies in many-electron atoms is

$$1s < 2s < 2p < 3s < 3p < 4s \approx 3d < 4p < 5s \approx 4d < 5p < 6s$$
$$\approx 5d \approx 4f < 6p < 7s \approx 6d \approx 5f < 7p$$

where the $\approx$ symbol means that orbitals differ little in energy and may compete for electrons. From the above sequence it is easy to predict at which element electrons start to occupy the s and p subshells. On the other hand the problem at which value of the atomic number electrons start to fill the d and f subshells requires the calculation of total energy of the atom in question for different electron configurations.

The order of filling shells and subshells by electrons follows two simple empirical rules:

– Subshells are filled by electrons in the order of increasing sum of quantum numbers $(n + l)$.

– For constant $(n + l)$, subshells are filled according to increasing quantum number n.

The rules are strictly obeyed by the Main Group elements, whereas the d and f electron elements sometimes show remarkable deviations. For instance, lanthanum has the configuration $5d^1 6s^2$ instead of $4f^1 6s^2$ and chromium $3d^5 4s^1$ instead of $3d^4 4s^2$. How the two rules act is shown in Fig. 2.1.

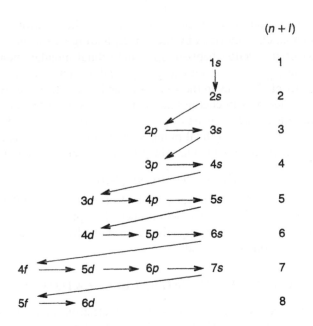

Fig. 2.1 The order of filling subshells by electrons.

In the sequence from hydrogen to argon the following subshells are filled consecutively :

$$1s, \quad 2s, \quad 2p, \quad 3s, \quad 3p$$

For $n = 3$ the $3d$ subshell is available. If the orbital energy were dependent only on the quantum number n, i.e. if $\varepsilon_{3d} < \varepsilon_{4s}$, then the electron configuration of potassium would be [Ar] $3d^1$. However, since there is a strong influence of the orbital momentum quantum number l, $\varepsilon_{4s} < \varepsilon_{3d}$ and the ground-state configuration of the K atom is [Ar] $4s^1$. The much greater radial probability density in the vicinity of the nucleus for the s than for the d electron is the reason why for potassium $\varepsilon_{4s} < \varepsilon_{3d}$. Frequent "excursions" of the s electron toward the nucleus make it more strongly bound and outweigh the effect of the principal quantum number. Similarly, in the case of Rb, Cs and Fr electrons fill the $5s$, $6s$ and $7s$ orbitals instead of $4d$, $5d$ and $6d$, respectively.

The $4s$ orbital is complete at calcium. In the scandium atom the next electron may occupy either the $3d$ or $4p$ subshell. Since the $4p$ orbital is the outermost orbital its energy depends, as in a hydrogen-like atom, almost exclusively on the principal quantum number. Thus $\varepsilon_{3d} < \varepsilon_{4p}$ and the electron configuration of the Sc atom is $[Ar]3d^1 4s^2$ and not $[Ar]4s^2 4p^1$. For the same reason electrons in yttrium, lanthanum and actinium occupy the $4d$, $5d$ and $6d$ and not the $5p$, $6p$ and $7p$ subshells, respectively.

2.2 EVOLUTION OF ORBITAL ENERGY

In connection with the competition of orbitals for electrons and the importance of this competition for the electron configuration of atoms it is interesting to study in more detail the evolution of orbital energies and radii as a function of Z, using the example of $3d$, $4s$ and $4p$ orbitals. Schematic changes of $3d$, $4s$ and $4p$ orbital energies are shown in Fig. 2.2.

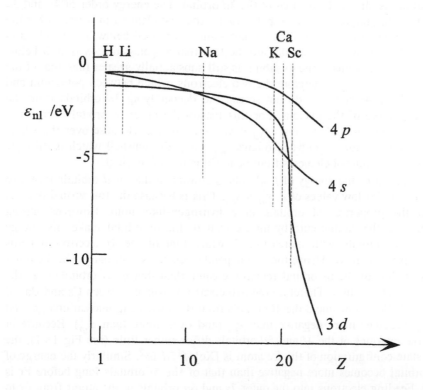

Fig. 2.2   Changes of $3d$, $4s$ and $4p$ orbital energies with Z from hydrogen to vanadium.

It can be seen in Fig. 2.2 that for low values of Z

$$\varepsilon_{3d} < \varepsilon_{4s}$$

which means that the orbital energy depends first of all on the principal quantum number n. The reason is that the virtual $3d$ and $4s$ orbitals show orbital radii much greater than the radii of orbitals which are filled across the second and third row. Therefore, the effective nuclear charge acting on these orbitals is almost equal to $+e$, because the $+Ze$ charge of the nucleus is almost completely screened by the $(Z-1)e$ charge of the remaining electrons. Hence, the properties of electrons in these orbitals would be very similar to those in hydrogen-like atoms i.e. depend only on the quantum number n. However, one should note that the radial probability density in the vicinity of the nucleus of the s electron is much greater than that of the d electron (eq. 1.11). Therefore, an electron if placed in the $4s$ orbital would be frequently subjected to the whole charge of the nucleus, which would make its orbital energy decrease with increasing Z very rapidly. On the other hand the $3d$ electron, which very rarely appears in the vicinity of the nucleus, would be much less sensitive towards the increasing nuclear charge, hence the $3d$ orbital energy remains almost constant, until approximately potassium. For this reason the two dependences cross very early (according to some calculations even before neon is reached). Indeed, one can see in Fig. 1.4 that in the excited Na atom the energy of the $4s$ orbital is already below that of the $3d$ orbital. The energy order of $4s$ and $3d$ orbitals becomes important only at potassium – the first element in which either the $3d$ or $4s$ orbital can be filled. Since at potassium $4s$ is well below $3d$, the electron enters the $4s$ orbital. However, despite the fact that at potassium $4s$ is well below $3d$, $\varepsilon_{4s} < \varepsilon_{3d}$, the radius of the $4s$ orbital is still substantially greater than that of the $3d$ orbital, $\langle r_{4s} \rangle > \langle r_{3d} \rangle$. Therefore the two electrons, which between potassium and calcium fill the $4s$ subshell, do not screen the deeper lying $3d$ orbitals from the increasing charge of the nucleus. For this reason the order of orbital energies is reversed at scandium, where now $\varepsilon_{3d} < \varepsilon_{4s}$, see Fig. 2.2. Because over the whole range of Z (except for $Z = 1$) we also have $\varepsilon_{4p} > \varepsilon_{4s}$, the subshell which is filled by electrons in the range of elements from Sc to Cu is the $3d$ subshell.

The evolution of the energy of f orbitals is similar to that of d orbitals (see also Section 18.3). For low values of Z, $\varepsilon_{4f} < \varepsilon_{6s}$. This is because the two virtual orbitals manifest the properties of orbitals in a hydrogen-like atom. However, strong penetration of the atomic core by an electron in the $6s$ orbital makes its energy decrease very rapidly with increasing Z, while that of the $4f$ electron remains constant up to about Xe. Hence the two dependences cross very early. On the other hand the radius of the $6s$ orbital remains greater than that of $4f$ orbitals over the whole range of Z values. Therefore the two electrons, which between Cs and Ba fill the $6s$ subshell, do not shield the $4f$ orbitals from the increasing nuclear charge, and at Ce ε_{4f} becomes more negative than ε_{6s} (and even more than ε_{6p}). Because at cerium the energies of the $4f$ and $5d$ orbitals differ rather little (see Fig. 18.1), the ground state configuration of the Ce atom is $[Xe]4f^1 5d^1 6s^2$. Similarly the energy of the $7s$ orbital becomes more negative than that of the $5f$ orbitals long before Fr is reached. Feeding electrons into the outer $7s$ and $6d$ orbitals in the atoms from Fr to Th markedly increases the effective nuclear charge acting on the deeper lying $5f$

orbitals. This makes the $5f$ orbital energies at Pa (configuration $[Rn]5f^26d^17s^2$) more negative than the $7s$ and $7p$. Because the $5f$ and $6d$ orbitals have similar energies (Fig. 18.2) they compete for electrons not only in Pa, U, and Np but also in Cm, which has the mixed $[Rn]5f^76d^17s^2$ configuration. Similarly, in the lanthanide series Gd has the mixed $[Xe]4f^75d^16s^2$ configuration, see Section 18.5.

2.3 OCCUPATION OF ORBITALS AND THEIR ENERGY

Since $\varepsilon_{3d} < \varepsilon_{4s}$ in the scandium atom the question arises why the ground state configuration of Sc is $[Ar]3d^14s^2$ rather than $[Ar]3d^3$, or at least $[Ar]3d^24s^1$. The same question concerns all d electron elements which have an unfilled $(n-1)d$ subshell and two (or one) electrons in the ns subshell. We know that the ground-state configuration is that which corresponds with the lowest total energy of the atom, E. Calculations by the Hartree method show that for scandium

$$E([Ar]3d^14s^2) < E([Ar]3d^24s^1) < E([Ar]3d^3)$$

This sequence of total energies can be explained qualitatively by the action of two opposite tendencies. On the one hand $\varepsilon_{3d} < \varepsilon_{4s}$, which promotes the $[Ar]3d^3$ configuration, on the other hand electrons tend to occupy different subshells, because then the inter-electronic repulsion is the lowest. The second factor favours both the $[Ar]3d^14s^2$ and the $[Ar]3d^24s^1$ configurations over the $[Ar]3d^3$ configuration. From the first two the $3d^14s^2$ configuration is energetically more favourable. The probable reason is that in the scandium atom the radial extent of the $4s$ orbital is much greater than that of the $3d$ orbitals, $\langle r_{4s} \rangle = 208$, while $\langle r_{3d} \rangle = 89$ pm. Because of smaller radius repulsion between two electrons in the $3d$ subshell is much greater than between two electrons in the $4s$ subshell which, of course, makes the $3d^24s^1$ configuration less stable.

Fig. 2.3 shows the energies of $3d$ and $4s$ orbitals in the Sc atom for the $3d^14s^2$ and $3d^24s^1$ configurations and the ns and $(n-1)d$ orbital energies in the ground states of V and Nb atoms. One can see in this Figure that on passing from d^1s^2 to d^2s^1 and from d^3s^2 to d^4s^1 configuration the d orbital energy increases considerably, that of the s orbital much less. The increase of the d orbital energy with addition of one electron is the result of increased total energy of repulsion between the d electrons. The small increase in the s orbital energy arises from two opposite tendencies. On the one hand the energy of the ns orbital becomes more negative, because with one electron in the s orbital there is no inter-electronic repulsion in this orbital. On the other hand the orbital energy of the s electron increases because of better shielding from the nuclear charge by the increased number of deeper lying d electrons. Apparently the second factor prevails and the energy of the ns orbital increases. However, a transfer of an electron from an ns to an $(n-1)d$ subshell takes place in Cr and Cu, which have the ground-state configurations $[Ar]3d^54s^1$ and $[Ar]3d^{10}4s^1$ instead of $[Ar]3d^44s^2$ and $[Ar]3d^94s^2$, respectively. Among $5d$ elements Pt and Au have the s^1 configuration, uncommon in this row. In the second transition series Pd even has the $[Kr]4d^{10}5s^0$ configuration. The reason for the transfer of the s electron to a d subshell is discussed in Sections 15.5 and 16.1.

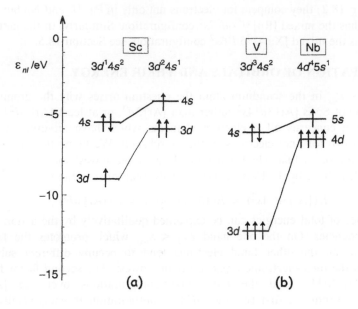

Fig. 2.3 The dependence of orbital energy on orbital occupation: a) different configurations of the scandium atom (orbital energies for the $3d^24s^1$ configuration are estimated values); b) comparison of vanadium with niobium.

3

Radii and their changes in the Main-Group elements

The radius of a subshell is the radius, $\langle r_{nl} \rangle$ or $r_{max,nl}$, of the orbital which belongs to this subshell. In the case of strong splitting of p, d and f orbitals, which occurs in heavy atoms, we use the radii of $p_{1/2}$, $p_{3/2}$, $d_{3/2}$, $d_{5/2}$, $f_{5/2}$ and $f_{7/2}$ orbitals weighted for closed subshells over their electron occupancy. As the radius of the atom, R, we adopt in this book the radius $\langle r_{nl} \rangle$ or $r_{max,nl}$ of the outermost subshell occupied by electrons. The radii of outer subshells determine the experimental covalent, metallic, ionic and van der Waals radii. The radius of the atom and the radii of outer subshells are very important characteristics, because their changes down the Groups affect changes of many chemical properties of respective elements. Although there is a close correlation between the radius of the orbital and its energy (except for the f and $6d$ orbitals), in this book we mainly use orbital radii to explain properties of elements, because of their more easily visualisable character.

3.1 SCREENING FROM THE NUCLEAR CHARGE

Since electrons screen the nuclear attraction, the positive charge acting on a specific electron is not equal to $+Ze$ but $+Z_{eff}e$, where Z_{eff} denotes the effective atomic number given by

$$Z_{eff} = Z - S \tag{3.1}$$

where S is the shielding constant calculated from the Slater rules. The Slater rules are as follows:

1. The electronic structure of the atom is written in groupings:

   (1s) (2s, 2p) (3s, 3p) (3d) (4s, 4p) (4d, 4f) and so on.

2. We then assume that:

   - Electrons in higher orbitals i.e. to the right in the above list, do not shield the nuclear charge for inner electrons, so that their contribution to S is nothing.

   - For ns and np valence electrons the contribution to S is:

     0.35 for each electron in the same grouping, except the $1s$ electron which contributes 0.3;

     0.85 for each electron in the $n-1$ grouping; or

     1 for each electron in the $n-2$ and lower groupings.

   - For nd and nf valence electrons the contribution from electrons in the same grouping is 0.35 and that from groupings to the left equals 1.0.

Slater's rules give only approximate values of shielding constants, mainly because penetration of core orbitals by valence electrons is only partly taken into account. Exact values can be obtained from quantum-mechanical calculations. It is interesting to notice the difference between shielding of the nuclear charge for s, p and for d, f electrons by inner electrons. As shown in the above listing, electrons in the $n-1$ grouping contribute 0.85 to shielding constants for s and p electrons and 1.0 for d and f electrons. The reason is that d and f electrons, in contrast to s and p electrons, only negligibly penetrate the vicinity of the nucleus (see eq. 1.11) and are, therefore, well screened from the nuclear charge by electrons in the next inner shell. Screening of s and p electrons by the next inner shell is less efficient because they have a higher probability to reside near the nucleus. As shown by the data in Table 3.1 the effective nuclear charge, from Slater's rules, acting on the valence electrons increases as the shells are filled by electrons. The increase of the effective charge across the Period results in a decrease of shell radii, and increase of ionization potentials and electronegativity. The Slater effective nuclear charge acting on the outermost electron increases down each Group, as far as the third row for s-elements and the fourth row for p-elements. Z_{eff} is constant down d-block Groups.

Table 3.1 – Effective nuclear charge, Z_{eff}, acting on the outermost s and p electrons, from Slater's rules.

Li	Be	B	C	N	O	F	Ne
1.30	1.95	2.60	3.25	3.90	4.55	5.20	5.85
Na	Mg	Al	Si	P	S	Cl	Ar
2.20	2.85	3.50	4.15	4.80	5.45	6.10	6.75

3.2 CHANGES OF ATOMIC RADII IN s AND p BLOCK ELEMENTS

The appearance of a new subshell and its filling with electrons affects the radius of the atom, R, in the following way:

1. The beginning of a new shell, i.e. the ns subshell results in a big increase of the atom radius.

2. The radius of the atom also increases when the first electron enters a p subshell, except for the 2p subshell.

3. The appearance of the first d and f electron does not increase the radius of the atom, because d and f orbitals are inner orbitals.

4. The radius of the atom decreases with filling both outer and inner orbitals by electrons, which is the result of incomplete shielding.

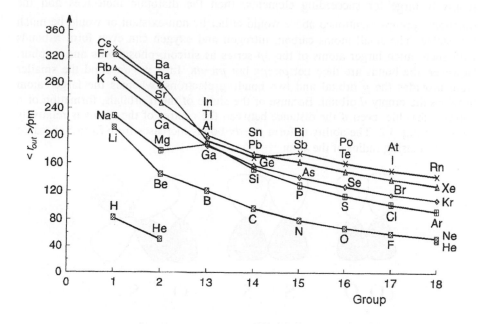

Fig. 3.1 The radius of the atom, $R = \langle r_{out} \rangle$, of s and p block elements. In accordance with its electron configuration helium has been placed also in Group 2.

Fig. 3.1 shows the radii of the atoms, equal to the radii of the outermost orbitals $R = \langle r_{out} \rangle$, for s and p block elements. Changes in the radii across Periods and Groups of the Main Group elements are as follows:

1. The radius R decreases between hydrogen and helium due to increasing Z_{eff} and increases significantly with occupying the $2s$ shell by the first electron at lithium. Due to incomplete shielding the radius R decreases between lithium and beryllium and again between beryllium and boron, in spite of adding the first electron to the $2p$ shell at boron. The reason is that the effect of increasing Z_{eff} prevails over a small increase due to the appearance of a new subshell. The increase of Z_{eff} across the $2p$ series makes the radii of atoms from boron to neon decrease significantly. The very small radii of B, C, N, O and F atoms are responsible for unique chemical properties of these elements. For instance, only carbon, nitrogen and oxygen form $p\pi-p\pi$ bonds in homo- and hetero-nuclear molecules such as

$$C = C \quad N \equiv N \quad O = O \quad C = O \quad N = O$$

and in the following functional groups:

$$> C = C < \quad -C \equiv C- \quad > C = O \quad -C \equiv N \quad -N = N-.$$

Moreover $p\pi-p\pi$ bonds also appear in aromatic rings, in graphite, and in fullerenes. Formation of $p\pi-p\pi$ bonds requires a small distance between atom centres – such bonding is feasible only between atoms of small radius, see Fig. 3.2. If atomic radii increased significantly between beryllium and boron and consequently were relatively large for succeeding elements, then the diatomic molecules and the functional groups mentioned above would either be non-existent or would be much less stable. The small atoms carbon, nitrogen and oxygen can even form π bonds with such much larger atoms of the $3p$ series as silicon, phosphorus and sulphur. However, the bonds are then not $p\pi-p\pi$ but $p\pi-d\pi$. In a $p\pi-d\pi$ bond the smaller atom provides the p orbital and two bonding electrons, whereas the larger atom provides the empty d orbital. Because of the shape of the d orbitals, formation of π bonds is feasible, even if the distance between the centres of the atoms is relatively large, see Fig. 3.2. The ability to form $p\pi-d\pi$ bonds increases from Si to S, because of the decreasing radius of the atom.

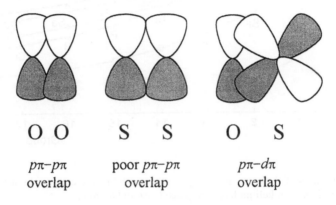

O O	S S	O S
$p\pi-p\pi$ overlap	poor $p\pi-p\pi$ overlap	$p\pi-d\pi$ overlap

Fig. 3.2 Correlation between atom size and the formation of π bonds.

2. In the third Period the radius of the atom decreases between Na and Mg and then increases considerably between Mg and Al, in contrast to the decrease between Be and B. The reason is the building up of the $3p$ subshell, which has a much greater radius than the $3s$ shell. The increase between Mg and Al results in much greater radii of $3p$ element atoms than of their lighter congeners. Therefore, the $3p$ elements only very rarely form π bonds between themselves and with each another. Examples of $p\pi-p\pi$ bond formation by $3p$ elements are bonding in the $S = S$ molecule and in the $\equiv P = S$ functional group. However, it should be noticed that recently one of the silicon alkenes $R_2Si = SiR_2$ containing a $p\pi-p\pi$ bond between the silicon atoms, which are considerably larger than sulphur atoms, has been isolated. Higher radii and lower ionization potentials make the $3p$ (and also $4p$ and $5p$) elements very different from their lighter congeners. The question arises why the $2p$ shell, which in contrast to higher p shells, is a nodeless shell, has an abnormally small size. The explanation which has been advanced is that the $2p$, like the $3d$ and $4f$ subshells, has no inner analogues to which it would have to be orthogonal – in other words it does not experience the so-called "primogenic repulsion". A figurative explanation is that p orbitals differ from s orbitals in angular charge distribution, hence repulsion between electrons in $2p$ and electrons in deeper lying $1s$ and $2s$ orbitals is weak and, consequently, the radius $\langle r_{2p} \rangle$ is relatively small. On the other hand the $3p$ orbitals (and also $4p$, $5p$ and $6p$) which have the same symmetry as the inner $2p$ ($3p$, $4p$ and

$5p$) orbitals would experience strong repulsion if they had not expanded. This effect is illustrated in Fig. 3.3, which shows a small difference in radial extent between $2s$ and $2p_{1/2}$ orbitals and much greater differences between ns and $np_{1/2}$ orbitals for $n \geq 3$ in Group 13 elements. In the same way (see Fig. 3.1) repulsion between the $2s$ and $1s$ electrons makes the lithium atom much larger than the hydrogen atom (the observed difference is the result of both building a new shell and interelectronic repulsion). One should also notice that when helium, in accordance with its electron configuration, is placed below beryllium in Group 2 then the difference in the radius R between the first and second element in Group 2 becomes large, as is the case with elements of Group 1 and Groups 13 to 18 asssuming that the first element in Group 18 is neon. We shall see in Section 15.6.2 that the radii of d orbitals follow the same pattern.

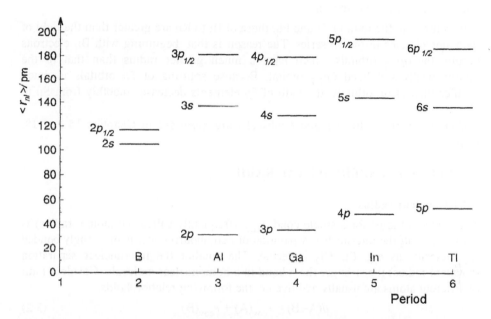

Fig. 3.3 The difference in radial extent between the $np_{1/2}$ orbitals and ns subshells in Group 13 elements.

3. Fig. 3.1 shows a big difference in radii between potassium and sodium atoms. The reason is that the $3s$ shell in Na is built over the small $2p$ shell of Ne, whereas the $4s$ shell in K is built over the considerably larger $3p$ shell of Ar. This is an example of how orbital properties of p block elements are transferred to s block elements.

4. Were it not for the build-up of the $3d$ subshell, the difference between the radii of gallium and aluminium atoms would be similar to that between potassium and sodium or calcium and magnesium. However, incomplete screening from the nuclear charge by $3d$ electrons makes the Ga radius even slightly smaller than that

of Al. Generally, the presence of a filled $3d$ shell makes the $4p$ elements more similar to their $3p$ congeners than to other elements in the Group.

5. In the absence of a filled $4f$ subshell the difference between the radii of thallium and indium atoms should be approximately equal to that between caesium and rubidium. This is because the effect on the radius R of filled $4d$ and $5d$ subshells and of building up the $5p$ and $6p$ subshells is very similar. However, incomplete screening from the nuclear charge by f electrons decreases the radius of thallium and the elements that follow. An additional factor is the relativistic effect (see Section 4.6) which decreases the radius of the outermost $6p_{1/2}$ orbital in Tl and Pb. Because of relativistic stabilization of the $6p_{1/2}$ orbital, which is occupied by one electron in Tl and two electrons in Pb, the radii of the Tl and Pb atoms are even smaller than the radii of the In and Sn atoms, respectively. Both factors, i.e. the presence of the filled f shell and the relativistic effect, result in remarkable similarities between Tl and In and between Pb and Sn.

In contrast to the radii of Tl and Pb, those of Bi to Rn are greater than the radii of their homologues in the $5p$ series. The reason is that, beginning with Bi, electrons occupy the $6p_{3/2}$ orbitals which have a much greater radius than that of the relativistically stabilized $6p_{1/2}$ orbital. Because splitting of $5p$ orbitals is much smaller than of $6p$ orbitals, the radii of $5p$ elements decrease smoothly from Sn to Xe.

Changes in the radii of d and f subshells are discussed in Chapters 15 and 18, respectively.

3.3 TYPES OF EXPERIMENTAL RADII

3.3.1 Covalent radius

The covalent radius for a single bond, r_{cov}, (frequently called the atomic radius) is defined as half the internuclear separation of neighbouring atoms in a singly bonded A_2 molecule as e.g. the Cl_2 molecule. The smaller the internuclear separation (covalent radius) the stronger the σ bond in a homonuclear molecule. Covalent radii of different atoms are usually additive i.e. the following relation holds:

$$d(A-B) = r_{cov}(A) + r_{cov}(B) \tag{3.2}$$

where $d(A-B)$ is the experimentally determined distance between the centres of the A and B atoms; $r_{cov}(A) = \frac{1}{2}d(A-A)$ and $r_{cov}(B) = \frac{1}{2}d(B-B)$.

The strength of a covalent bond should be at maximum when the internuclear distance between the two bonded atoms is such that the maximum of radial probability density of one atom overlaps that of the other. Fig. 3.4 shows the dependence of r_{cov} on the radius of maximum radial probability density for the outermost orbital, $r_{max,out}$. Fig. 3.4 shows that, with the notable exception of hydrogen:

  – r_{cov} is a relatively good linear function of $r_{max,out}$;

  – r_{cov} is, as expected, very similar to $r_{max,out}$

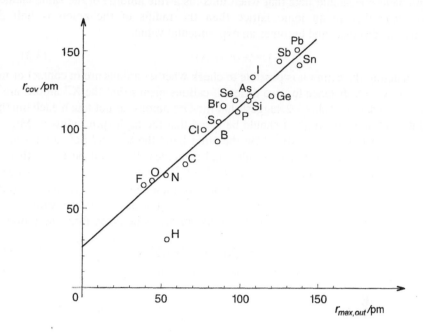

Fig. 3.4 Dependence of the covalent radius, r_{cov}, on the radius of the outermost shell in the atom, $r_{max,out}$.

3.3.2 Metallic radius

The metallic radius, r_{met}, of an element is defined as equal to half the experimentally determined distance between the nuclei of nearest neighbour atoms in a metallic solid. The smaller the metallic radius the stronger is metallic bonding as measured by atomization energy and, approximately, by melting or boiling temperatures. Since metallic bonding is a highly delocalized form of covalent (electron-deficient) bonding one can also expect a correlation between the internuclear distance and $r_{max,out}$. Fig. 3.5 shows the dependence of r_{met} on $r_{max,out}$ for a number of metallic elements. One can see that:

- r_{met} depends linearly on $r_{max,out}$
- r_{met} is, as expected, approximately equal to $r_{max,out}$.

Comparison of Fig. 3.5 with Fig. 3.4 shows almost the same slopes for the two plots, equal to about 0.87. This suggests that with increasing radius of the atom the strengths of both covalent and metallic bonding decrease at the same rate.

3.3.3 Ionic radius

From X-ray or neutron diffraction studies we know only the distance between the nuclei of neighbouring cations and anions. The difficulty lies in dividing such distances into a contribution from the cation, r_+, and from the anion, r_-. To tackle this problem one should note that when ions (as a rule anions) of the same element touch each other in an ionic lattice then the radius of the anion is half the internuclear distance, and becomes an experimental value:

$$r_i(A^-) = \tfrac{1}{2}\,d(A-A) \tag{3.3}$$

By changing the cation it is possible to check whether anions are in contact or not. The same $M\cdots X$ distance for two different cations means that the X^- anions are in contact. The data in Table 3.2 suggest that the O^{2-} anions do not touch each another in MnO, but do so in MgO. It should be noted that the high-spin radius of Mn^{2+} is 11 pm greater than that of Mg^{2+}. On the other hand the S^{2-} and Se^{2-} anions are in close contact in the respective sulfides and selenides. It should be noted that the $r_i(2+)$ radii of Mg and Mn are very close. The O^{2-} anions are in contact in silicates and in spinels. The $O\cdots O$ distance measured in these compounds yields an $r_i(O^{2-})$ radius equal to 140 pm. Similar measurements on fluorides result in a value of 133 pm for the radius $r_i(F^-)$. The radii of cations are then calculated from the following relations:

$$r_i(M^{z+}) = d(M\cdots O) - r_i(O^{2-}) \tag{3.4}$$

$$r_i(M^{z+}) = d(M\cdots F) - r_i(F^-) \tag{3.5}$$

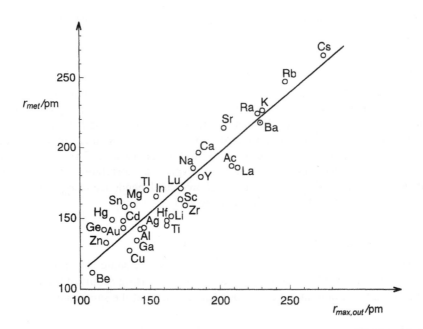

Fig. 3.5 Dependence of the metallic radius, r_{met}, on the radius of the outermost shell in the atom, $r_{max,out}$.

In this method we assume that the $r_i(O^{2-})$ radius calculated from the $O \cdots O$ distance is the same in all ionic salts i.e. is independent of the cation. We also assume that the radius of the cation calculated from the $M \cdots O$ distance retains its value in other ionic compounds e.g. in chlorides. These additivity rules hold well only for ions of low charge and large radius. However, the radius of the cation strongly depends on the coordination number, CN, Table 3.3.

Table 3.2 – The effect of the cation on the $M \cdots X$ distance, in pm.

MgO	MnO	MgS	MnS	MgSe	MnSe
210	224	260	259	273	273

Table 3.3 – Variation of ionic radii of cations (in pm) with coordination number.

CN	Sr^{2+}	Zn^{2+}	In^{3+}	Zr^{4+}
4	–	60	62	59
6	118	74	80	72
8	126	90	92	84

Despite the doubts and difficulties attendant on the establishment of ionic radii, it should be emphasised that values on the various empirical scales show remarkably close agreement. The only marked exception is provided by Li^+, for which values range from 60 to 78 pm. This variation may arise at least in part from uncertainties as to whether it is anion-anion (as in LiI) or anion-cation contact which determines dimensions in various salts LiX.

There is a very good linear dependence of the cationic radius of the elements of Groups 1, 2, 12, 13 and 14 on the radius of the outermost shell of the respective cations, $\langle r_{nl,out} \rangle$, Fig. 3.6. It can be seen in this Figure that:

- for each charge on the cation there is a separate linear relation;
- for the same value of $\langle r_{nl,out} \rangle$ the radius of the cation decreases with increasing charge;
- the radius of the cation is as a rule greater than the radius of its outer-most shell. However, the difference between r_i and $\langle r_{nl,out} \rangle$ decreases with increasing charge on the cation;

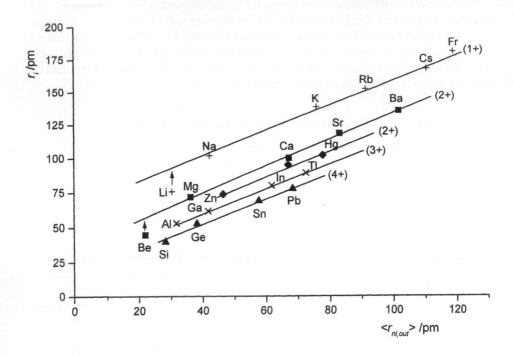

Fig. 3.6 Dependence of the ionic radius, r_i (CN 6), of Group 1, 2, 13 and 14 elements on the radius of the outermost shell in the ion, $\langle r_{nl,\text{out}} \rangle$.

- Li^+ and Be^{2+} show remarkable deviations, perhaps attributable to their respective outer shells differing from those in their heavier congeners (s in contrast to p subshell). Another explanation is the small radius of the outermost shell in the ion which is 30.3 pm for Li^+ and 21.9 pm for Be^{2+}. Such small radii result in shifting electrons from the anion to the cation, which confers some covalent character on the bonding. Covalence decreases the experimental $d(M \cdots O)$ distance, hence the calculated radii of these cations, eq. 3.4.

Similar linear dependences are also shown by cations of d- and f-block elements (see Figs. 15.3, 15.4 and 18.5).

The estimation of "absolute" ionic radii from electron density maps, long after the establishment of the internally consistent and closely comparable sets of empirical ionic radii discussed above, complicated the situation somewhat. The electron density maps, obtained for such salts as LiF, NaCl, KCl, MgO, and CaF_2, indicated the empirical ionic radii to be systematically in error by approximately 15 pm – cations being too small, anions too large, by this amount. A comprehensive set of *crystal radii* was therefore established to take into account the electron density results, it being suggested that crystal radii referred to ions in the crystal lattice, ionic radii to "free ions". The relatively small differences between crystal and ionic radii have a negligible effect on correlations and discussions, except for those involving such quantities as radius ratios where the differences are cumulative. In the present book we use ionic radii throughout, with all values taken from Shannon's extensive tabulation (cf. Preface) where, incidentally, both ionic and crystal radii are listed.

3.3.4 Van der Waals radius

The van der Waals volume, V_w, is the volume of an atom or a molecule which is not accessible to other atoms (molecules) because of repulsion between nonbonding shells. The energy of repulsion between shells is the consequence of the Pauli principle and strongly decreases with decreasing distance between the centres of the atoms:

$$E_{rep} = B / d^{12} \tag{3.6}$$

where B is a constant. Because of the strong dependence of the repulsion energy on the distance, atoms can be considered hard spheres. Atoms which are not bonded chemically attract each other by van der Waals forces so that the total energy, E, becomes:

$$E = B / d^{12} - A / d^6 \tag{3.7}$$

Fig. 3.7 shows how E depends on the internuclear distance. For atoms of the same element half of the distance d_0 is conventionally called the van der Waals radius:

$$r_W = \tfrac{1}{2} d_0 \tag{3.8}$$

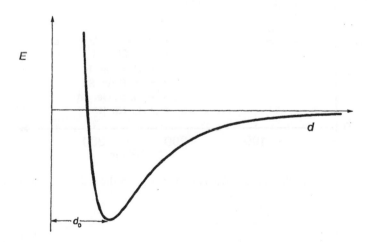

Fig. 3.7 Dependence of energy on the distance between two atoms which do not interact chemically.

Since for internuclear distances shorter than d_0 the repulsion energy is a very steep function of d, the van der Waals radius corresponds with very low radial probability density. The r_w values are best obtained from X–X contact distances in a solid phase formed by molecules containing chemically bonded X atoms. As seen in Fig. 3.8, r_w increases with increasing radius of the atom and is larger than R. Comparison of the data in Figs. 3.4, 3.5, 3.6 and 3.8 shows that overall the various radii decrease in the following order:

$$r_w > R > r_{met} > r_{cov} > r_i .$$

This order is valid for elements which form metallic phases and cations. In the case of elements which form anions, the ionic radius is much greater than the covalent and is slightly greater than the van der Waals radius. For instance, in the case of bromine the radii r_i, r_w, r_{cov}, and R are 196, 185, 114, and 112 pm respectively. It is interesting to note that the ionic radius of a halide ion X^- is similar to the van der Waals radius of the isoelectronic noble gas atom. For instance, $r_i(Br^-) = 196$ pm whereas $r_w(Kr) = 202$ pm.

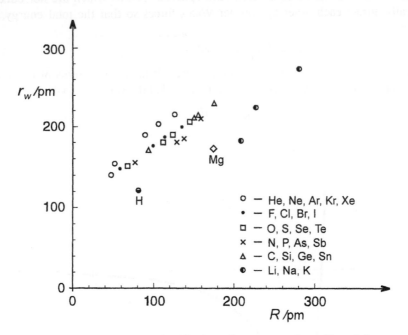

Fig. 3.8 Dependence of the van der Waals radius, r_w, on the radius of the atom, R.

4

Orbital energies and related properties

4.1 ORBITAL ENERGIES

Fig. 4.1 shows the energies of the outermost orbitals in atoms, ε_{out}, of s and p block elements. It follows from comparison of Fig. 4.1 with Fig. 3.1 that changes in orbital energies more or less closely follow changes in the radii of atoms $R = \langle r_{out} \rangle$. In particular:

- Building a new shell (the ns subshell) makes the orbital energy increase (its absolute value decreases). However, the change is less marked than that in the case of the radius of the atom, R. The largest increase in ε_{out} is observed between hydrogen and lithium. If according to its electron configuration helium is placed in Group 2 then the largest increase in ε_{out} is that between He and Be.
- Filling a subshell decreases ε_{out} (the absolute value increases) because of incomplete shielding from the nuclear charge.

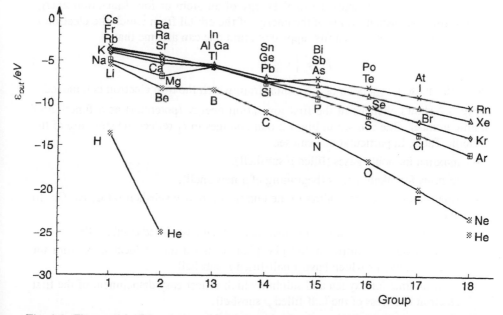

Fig. 4.1 The energy of the outermost orbital, ε_{out}, of s and p block elements. In accordance with its electron configuration helium has also been placed in Group 2.

- Equal values of ε_{out} for beryllium and boron, together with the increase of ε_{out} between magnesium and aluminum, make the $2p$ elements very different from their heavier congeners.
- For homologues in p block Groups the presence of the filled $3d$ shell brings the ε_{4p} orbital energy close to the ε_{3p} orbital energy.
- The presence of the filled $4f$ shell and relativistic effects slightly decrease the gap between the ε_{5p} and ε_{6p} orbital energies for homologues in p block Groups. Just as in the case of radii, and for the same reason, the positions of thallium/indium and of tin/lead are reversed.

4.2 IONIZATION ENERGIES

Ionization energy or ionization potential (in kJ mol^{-1} or eV per atom) is the energy necessary to detach an electron from a gaseous atom. This parameter is an experimentally accessible quantity.

$$M^0(g) \rightarrow M^+(g) + e^- \qquad I_1 \tag{4.1}$$

$$M^+(g) \rightarrow M^{2+}(g) + e^- \qquad I_2 \quad \text{etc} \tag{4.2}$$

$$I_1 = E\,[M^+(g)] - E\,[M^0(g)] \tag{4.3}$$

$$I_2 = E\,[M^{2+}(g)] - E\,[M^+(g)] \qquad \text{etc} \tag{4.4}$$

In these equations E denotes the total energy of an atom or ion. Ionization energy differs from the absolute value of the energy of the orbital from which the electron is detached, ε_{out}. However, as a first approximation one can assume that

$$I_n \approx \left| \varepsilon_{out} \right| \tag{4.5}$$

where the subscript $_{out}$ denotes the orbital from which the n-th electron is removed.

Fig. 4.2 shows changes in the first ionization energy (potential) as a function of atomic number. One can see in Fig. 4.2 that changes in I_1 reflect the structure of the Periodic Table. In particular one can see:

- maxima for noble gases (filled p subshell);
- minima for alkali metals (beginning of a new shell);
- small minima corresponding to the outset of a new p subshell (at B, Al, Ga, In and Tl);
- maxima for zinc, cadmium and mercury, which have filled d subshells;
- local maxima for nitrogen and phosphorus, and a minor local maximum for arsenic, elements which have a half-filled p subshell;
- local minima for oxygen and sulphur which reflect easy detachment of the first electron in excess of the half-filled p subshell;
- the increase in I_1 across each row, because of incomplete shielding;
- the decrease in I_1 down each Group (parallel to increasing radius of the atom and orbital energy);

– homologous elements in the $2p$ and $3p$ rows show very different values of I_1, which reflects the difference in the radius and in the orbital energy ε_{np}.

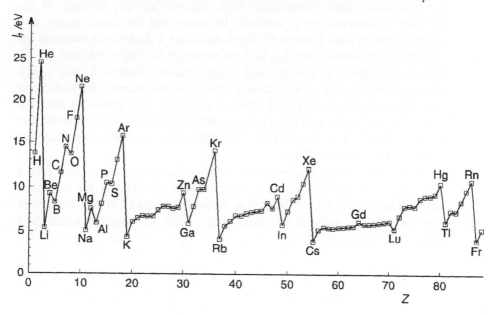

Fig. 4.2   Dependence of the first ionization potential, I_1, on the atomic number, Z.

The exceptionally high value of I_1 for mercury is the result of the direct relativistic effect, see Section 4.6. The reason for the minima at oxygen and sulphur is that the electron in excess of the half-filled p subshell must have its spin opposed to the other three. This additionally increases interelectronic repulsion (compare Sections 15.5 and 18.5). For the same reason I_2 shows a local minimum at fluorine.

4.3 ELECTRON AFFINITY

Chemical properties of elements, in particular of non-metals, depend on the electron affinity, A. This is defined as the energy which is released in the process of attaching an electron to a gaseous atom or ion:

$$X (g) + e^- \rightarrow X^- (g) \tag{4.6}$$

In the case of an M^{n+} cation the electron affinity is equal to the ionization energy I_n.

Electron affinity may be positive (easy attachment of an electron) or negative (the system with an added electron is unstable). Fig. 4.3 depicts how the electron affinity of atoms changes in the Periodic Table, showing the following main features.

– For p- and d-block elements electron affinities increase in each row from left to right (more gradually for d-block elements). The increase is due to increasing effective charge which, in turn, results from incomplete screening.

– Attachment of an electron to an atom of a noble gas or a Group 2 or 12 element is accompanied by a negative or near zero affinity, because it involves the start of a new shell or subshell without increase in the charge on the nucleus.

- Halogen atoms show particularly high electron affinity, because of the tendency to complete the p subshell. Hydrogen and the alkali metals also exhibit relatively high electron affinities, reflecting a tendency to complete the s subshell. The reason is that Z_{eff} is at its maximum for a filled subshell.

- The electron affinity of nitrogen and of phosphorus, which both have a half-filled p subshell, is low. A low electron affinity is also observed for manganese and rhenium, which have half-filled d subshells. The reason is that the electron in excess of a half-filled p or d subshell must have its spin opposed to the other three or five, respectively. This additionally increases interelectronic repulsion and decreases affinity.

- Local maxima in electron affinities are observed at copper, silver, and gold, a consequence of the tendency to complete the s subshell, as for the alkali metals. The electron affinity of gold is particularly high (2.31 eV), because of relativistic effects and the presence of the filled $4f$ subshell, which both stabilize the $6s$ orbital.

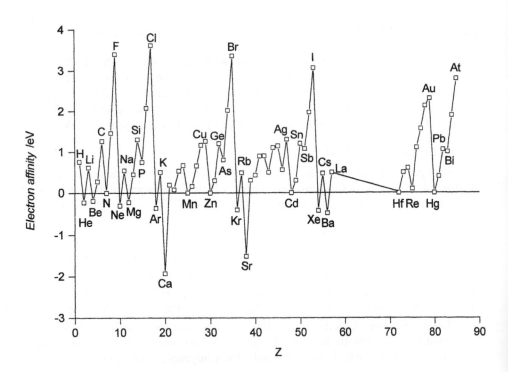

Fig. 4.3 Dependence of the electron affinity, A, on the atomic number, Z.

4.4 ELECTRONEGATIVITY

According to one qualitative definition electronegativity, χ, is the power of an atom **in a molecule** to attract electrons. There are several methods to assign numerical estimates to electronegativities. Each of these methods is based on a different quantitative approach and emphasises a different aspect of electronegativity.

The method of Pauling. The energy of an X–Y bond usually exceeds the mean of the X–X and Y–Y bond energies. If the electronegativity of Y is greater than that of X, the excess energy, Δ, can be attributed to the ionic-covalent resonance

$$X^+Y^- \leftrightarrow X{-}Y,$$

and can serve as a measure of the difference in electronegativities between Y and X. According to Pauling

$$(\chi_X - \chi_Y)^2 = \Delta/96 \tag{4.7}$$

where

$$\Delta = E(X{-}Y) - \tfrac{1}{2}\{E(X{-}X) + E(Y{-}Y)\} \tag{4.8}$$

and bond energies are in kJ mol^{-1}. Using a wide range of bond energies the χ_X and χ_Y values can be determined separately. Pauling electronegativities are then calculated assuming that the electronegativity of hydrogen is 2.20.

The method of Mulliken. The tendency of an atom in a molecule to compete with another atom for an electron should be proportional to the average of the ionization potential and electron affinity:

$$\chi \propto \tfrac{1}{2}(I + A) \tag{4.9}$$

The thinking underlying this definition is as follows. If an atom has a high ionization potential and a high electron affinity then it will easily acquire electrons but be reluctant to lose them. It will then show high electronegativity. On the other hand when both I and A are low, the atom will easily lose electrons but attach them only with difficulty. Hence it will have a low value of χ and would be classified as electropositive.

The method of Allred and Rochow. Electronegativity represents the force exerted by one of the nuclei forming the bond on the bonding electron.

$$\chi \propto Z_{eff}e^2 / r_{cov}^2 \tag{4.10}$$

where r_{cov} denotes the covalent radius.

When appropriately adjusted the three methods yield similar results. In this book we use Pauling electronegativities. Changes of electronegativity with Z are shown in Fig. 4.4. Because of decreasing shielding from the nuclear charge and hence increasing effective potential, electronegativity increases in each Period as the s and p subshells become filled. In the Main Groups electronegativity usually decreases down each Group. Halogens and elements of the $2p$ series show particularly high values of electronegativity. The electronegativity of d block elements initially increases in each series and then becomes almost constant. Because of relativistic effects gold shows very high electronegativity ($\chi = 2.54$). That is the reason why gold shows oxidation number −1 and forms with caesium ($\chi = 0.79$) a salt-like compound Cs^+Au^-.

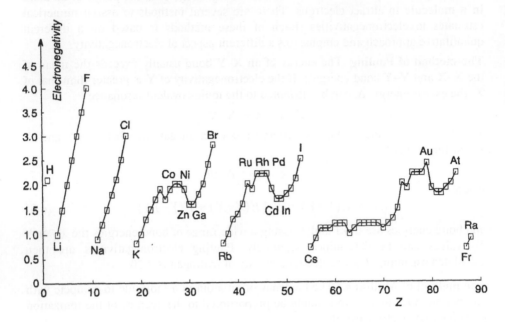

Fig. 4.4   Dependence of Pauling electronegativities on atomic number, Z.

In general, highly electronegative elements react eagerly with highly electropositive elements and *vice versa*. Bonding then shows ionic character. Elements with intermediate electronegativity react readily with both more and less electronegative elements, and bonding is then mainly covalent.

4.5 HARDNESS AND SOFTNESS

According to Mullikan's definition the *mean* of the ionization energy and electron affinity gives the electronegativity of an atom in a molecule. On the other hand half of the *difference* between the two energies in eV can serve as a measure of the so-called chemical hardness, η, of the atom, ion or molecule:

$$\eta = \tfrac{1}{2}(I - A) \tag{4.11}$$

According to eq. (4.11) hard atoms are those with high ionization energies and low electron affinities. Because the first ionization energy of an atom is as a rule much greater than its electron affinity, the hardest atoms are small atoms with high I_1 values, such as fluorine, oxygen and nitrogen which have η equal to 7.0, 6.1 and 7.3, respectively. In contrast, the alkali metals with their low I_1 and moderate A values are the softest elements with η ranging from 2.35 for lithium to 1.7 for caesium. Also lanthanides are soft, having hardness values of about 2.8. The hardness of most p and d block elements is intermediate and is in the range from 3 to 5. The hardness of atoms decreases down a Group as I_1 decreases, as shown for instance by comparison of lithium with caesium.

Hardness and softness, the latter defined as the reciprocal of η, are closely related to another property of atoms and ions called polarizability, α. Polarizability is the proportionality coefficient in the correlation between the induced electric moment μ and the electric field E acting on the atom or ion:

$$\mu = \alpha E \qquad (4.12)$$

Polarizability measures the ease with which atom, ion or molecule is distorted by the external electric field, i.e. it measures what a mechanical model suggests as softness. Therefore, soft atoms and ions (low values of η) should show high values of α and *vice versa*. The following argument correlates polarizability more rigorously with softness. According to the quantum theory treatment of polarization, the wave function of the polarized state results from mixing the ground state function with suitable excited state wave functions. Both intuition and rigorous treatment tell us that such mixing is the more effective, hence polarizability greater, the smaller the gap between the two energies. Because the absolute values of the electron energy in the ground state and in the lowest-lying excited state are approximately equal to ionization energy and electron affinity, respectively, the correlation between softness and polarizability becomes evident.

Softness and hardness of ions, i.e. the ease (or lack of it) with which they are polarized by counterions, is of paramount importance in complex formation and in determining stability constants. Table 4.1 documents the hardness of some cations. It is seen in this Table that Group 1 and 2 cations, in particular the lightest members, show very high hardness. Most p and d block cations with the same charge show similar hardness, except for Al^{3+} and to some extent also Ga^{3+}. The very high hardness of Al^{3+} results from the fact that the fourth electron is detached from the very low-lying $2p$ subshell (see Section 3.2). In the case of gallium the fourth electron is also removed from the relatively low-lying $3d$ subshell. Lanthanide M^{3+} cations show moderate hardness, ranging from 8.3 for cerium to 9.3 for ytterbium.

Table 4.1 – Hardness (η) of some M^+, M^{2+} and M^{3+} cations, in eV

Li^+	35.1	Mg^{2+}	32.5	Al^{3+}	45.7
K^+	13.6	Ca^{2+}	19.5	Cr^{3+}	9.1
Cu^+	6.3	Fe^{2+}	7.2	Fe^{3+}	12.1
Cs^+	10.6	Zn^{2+}	10.9	Ga^{3+}	16.8
Tl^+	7.2	Sn^{2+}	7.9	Bi^{3+}	9.9

4.6 RELATIVISTIC EFFECTS AND PROPERTIES OF ELEMENTS

Relativistic effects influence many chemical properties of elements by changing orbital energies and radii, ionization potentials, electron affinities and electronegativities. Relativistic effects are already important for elements of the sixth row, in particular at the end of the $5d$ series, and for $6p$ elements. They are of paramount importance in the chemistry of the actinides and transactinides.

The so-called direct relativistic effect originates from the fact that no body can move with a speed greater than the speed of light ($c = 2.998 \times 10^8$ m s^{-1}). When this velocity is approached the mass of the body increases according to the equation

$$m = \frac{m_0}{\sqrt{1-(v/c)^2}} \tag{4.13}$$

where m_0 denotes the rest mass and v the velocity of the electron. The increase in the mass increases the kinetic energy of the electron, T, because the relativistic kinetic energy is given by the expression

$$T = m_0 c^2 (\gamma - 1) \tag{4.14}$$

where

$$\gamma = \frac{1}{\sqrt{1-(v/c)^2}}. \tag{4.15}$$

In the Coulomb field which dominates in the vicinity of the nucleus the average kinetic energy, T (a positive quantity), the total energy, E (a negative quantity), and the average potential energy, V (a negative quantity), are related to one another in the following manner

$$E = \overline{V} + \overline{T} \tag{4.16}$$

$$\overline{V} = -2\overline{T} \tag{4.17}$$

$$E = -\overline{T} \tag{4.18}$$

Because the increase in kinetic energy is accompanied by twice as big an increase in the bonding (negative) potential energy, the overall effect is an increase of the absolute value of the total energy, E. Thus, the relativistic effect makes the electron more strongly bound and decreases the corresponding orbital radius. Calculations show that in the case of gold the average velocity of the $1s$ electron is 0.58 that of light, the mass of the electron increases by 23%, the ε_{1s} orbital becomes more negative by 10%, and the orbital radius $\langle r_{1s} \rangle$ has shrunk by 12%.

The mass of the electron, its kinetic energy and the absolute value of its total energy increase the more, the greater is the probability for an electron to approach the nucleus. We know that the radial probability density for finding an electron near to the nucleus depends on the orbital quantum number l and is given by

$$P_{nl}(r) \propto (r/a_0)^{2l} r^2 \qquad (4.19)$$

Therefore, the probability of finding an electron in the vicinity of the nucleus decreases in the order $s > p > d > f$. Because the probability is particularly high for the s electron ($l = 0$) the direct relativistic effect is, firstly, the stabilization of s electrons. This stabilization affects not only $1s$ electrons but also all other s electrons including valence s electrons, which is of primary importance for the chemical properties of heavy elements. The second relativistic effect is the spin-orbit splitting of p, d and f orbitals into $j = l \pm \frac{1}{2}$, i.e. into lower-lying $p_{1/2}$, $d_{3/2}$, $f_{5/2}$ and higher-lying $p_{3/2}$, $d_{5/2}$ and $f_{7/2}$ orbitals. The spin-orbit splitting is a relativistic effect, because its magnitude is inversely proportional to c^2, hence it vanishes as c increases to infinity. With regard to spin-orbit splitting it should be noted that the $p_{1/2}$ electron has the same quantum number $j = \frac{1}{2}$ as the s electron. Therefore, it has high probability density in the vicinity of the nucleus and for the same quantum number n is as strongly stabilized by the direct effect as the s electron.

Stabilization of s and $p_{1/2}$ electrons generates the third effect called indirect relativistic effect which consists in destabilization of f and outer d orbitals, and is particularly evident for electrons in the $d_{5/2}$ and $f_{7/2}$ orbitals. The outer $p_{3/2}$ electrons are also destabilized. The reason for destabilization is that due to the direct relativistic effect in heavy elements the inner s and $p_{1/2}$ electrons are, on the average, closer to the nucleus than in the non-relativistic case. Therefore, they shield the outer d and f electrons more effectively from the nuclear charge, which makes the orbital energies of the latter less negative and the orbital radii greater. Because the relativistically expanded d and f orbitals screen the nuclear charge less effectively than the non-relativistic orbitals, the valence s and $p_{1/2}$ orbitals become even more tightly bound than expected from the direct effect. For instance, the $\langle r_{6s} \rangle$ radius in gold shrinks by as much as 17% and the ε_{6s} orbital energy becomes more negative by 32%, compared with 12% and 10%, respectively, for the $1s$ electron. The very high relativistic effect for $6s$ in gold (and also in platinum) is, in part, caused by the $d^{n+1}s^1$ configuration, unique among $5d$ elements, compared with the normal $d^n s^2$ configuration.

All three relativistic effects, i.e. direct, indirect and spin-orbit splitting, increase approximately as Z^2. One can, therefore, expect the stabilization of s and $p_{1/2}$ electrons first of all for the heaviest element in a given Group. This is illustrated by changes in the first ionization potential in some Groups of the Periodic Table. One can see from the data in Table 4.2 that I_1 decreases down a Group except for the last element, which has a higher value of I_1. Because destabilization of d and f orbitals also increases with the atomic number, it becomes particularly high for $5d$ electrons and is higher for $5f$ than for $4f$ electrons.

Stabilization of s electrons affects both electron configurations of atoms and chemical properties of heavy elements. Because of relativistic stabilization of $6s$ electrons and the presence of the $4f$ shell, the $5d$ elements have, in general, the $5d^n 6s^2$ configuration, contrary to $4d$ elements, most of which have the $4d^{n+1} 5s^1$ configuration.

Table 4.2 – Changes of I_1 (eV) in some Groups of the Periodic Table

Configuration					
s^1	s^2	$d^{10}s^1$	$d^{10}s^2$	s^2p^1	s^2p^2
K	Ca	Cu	Zn	Ga	Ge
4.34	6.11	7.72	9.39	6.00	7.90
Rb	Sr	Ag	Cd	In	Sn
4.18	5.69	7.57	8.99	5.78	7.34
Cs	Ba	Au	Hg	Tl	Pb
3.89	5.21	9.22	10.43	6.11	7.41
Fr	Ra				
4.15	5.28				

Stabilization of s electrons is responsible for many unusual properties of heavy elements as e.g.:

- the very noble character and the yellow colour of gold;
- formation of the −1 oxidation state by gold in Cs^+Au^-;
- the stability of the Au_2 molecule and of the isoelectronic Hg_2^{2+} cation;
- the low melting and boiling temperatures of mercury;
- the instability of thallium and lead in oxidation states +3 and +4, respectively.

Because of stabilization of the $p_{1/2}$ electrons, bismuth forms the oxidation state +1 and polonium oxidation state +2. The reason is that both ions have the $6s^26p_{1/2}^2$ configuration, which makes the removal of successive electrons from the relatively stable $p_{1/2}$ orbital somewhat difficult.

Destabilization of $5d_{5/2}$ orbitals results in:

- remarkable stability of the oxidation state +3 for Au, in contrast to the low stability of this oxidation state for Ag and Cu;
- greater stability of the oxidation state +4 for Pt than for Pd;
- formation of the oxidation state +6 by platinum, which has the configuration $[Xe]4f^{14}5d_{3/2}^45d_{5/2}^56s^1$.

Greater destabilization of $5f$ than of $4f$ orbitals (the effect of increasing Z) is one of the reasons why uranium, neptunium and plutonium form stable oxidation states higher than +3, in contrast to their neodymium, promethium and samarium homologues, for which the stable oxidation state is +3.

5

Oxidation states and their stability

5.1 VALENCE, HYPERVALENCE AND OXIDATION STATE

Valence is the most characteristic chemical property of an element. According to classical definition valence gives the number of hydrogen atoms that combine with the atom of an element in a binary hydride, or is equal to twice the number of oxygen atoms in the oxide. When combined with halogen atoms or radicals such as the phenyl group, Ph, the element shows valence equal to the number of substituents accommodated around the central atom. For instance, phosphorus is pentavalent in PF_5 and Ph_5P, and sulphur is hexavalent in SF_6. These and many other compounds of p block elements violate the octet rule that permits only eight electrons to be accommodated in the valence shell of the central atom. In principle there are three ways to hold the extra electrons beyond the octet – two electrons in PF_5 and in Ph_5P, four in SF_6. These three ways are detailed in the following paragraphs.

The first is expansion of the valence shell by using higher lying nd orbitals to make sp^3d or sp^3d^2 hybrid orbitals. However, this way of accommodating extra electrons is difficult to accept for light elements. The reason is that for light p elements the energy gap between $n(s,p)$ and nd orbitals is probably too large to make hybrid orbitals.

Another possibility is formation of electron-rich three-centre four-electron, 3c-4e, bonds. When two atoms combine, each of them supplying one orbital, e.g. the p_z orbital, two molecular orbitals – one bonding and one antibonding – are formed. When each atom donates one electron then only the bonding orbital is filled and the bond is a two-centre two-electron, 2c-2e, bond, Fig. 5.1(a). Three orbitals from three atoms yield three molecular orbitals: one bonding, one antibonding, and one nonbonding. When the total number of electrons is two, only the bonding orbital is filled and the bond is a three-centre two-electron, 3c-2e, bond, Fig. 5.1(b). With four electrons the nonbonding orbital is also filled and the bond is then designated a three-centre four-electron, 3c-4e, bond Fig. 5.1(c). The 3c-2e bond is electron-deficient, while the 3c-4e bond is electron-rich. Having this in mind we can now explain the bonding in PF_5 and similar molecules. The PF_5 molecule, whose structure is a trigonal bipyramid, has three equatorial bonds formed by the s, p_x and p_y orbitals of phosphorus and three p_z orbitals from three fluorine atoms. The F–P–F axial bond is a 3c-4e bond and is formed by the p_z orbital of the P atom and two p_z orbitals of the two apical F atoms. The 3c-4e bond holds the two extra electrons in the nonbonding orbital which belongs to the two apical fluorine atoms and is not shared with the phosphorus atom. In this way the extra electrons are removed from the valence shell of the phosphorus atom and the octet rule is preserved.

There is also another explanation which is best illustrated by the example of SF_6, Fig. 5.2. We can assume that the s orbital and the three p orbitals of the central S atom in SF_6 combine with four p_z orbitals from four F atoms to form four bonding

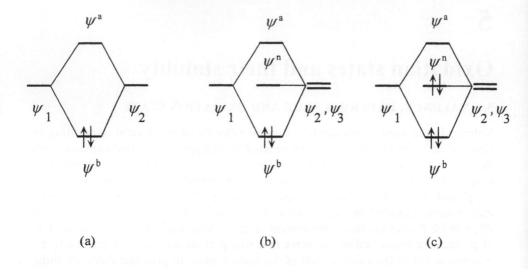

Fig. 5.1 (a) 2c-2e bond; (b) 3c-2e bond; (c) 3c-4e bond. The a, b and n superscripts denote antibonding, bonding and nonbonding orbitals, respectively.

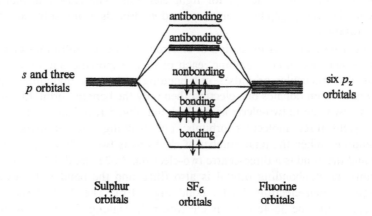

Fig. 5.2 Molecular orbitals in SF_6.

and four antibonding orbitals. The two remaining p_z orbitals from the two F atoms form two nonbonding molecular orbitals which belong only to the fluorine atoms and hold the four extra electrons. Compounds like PF_5 and SF_6 are called hypervalent compounds and are very common among p block elements.

In connection with valence one should also mention the concept of valence state. This is a hypothetical state of the atom in which it is ready to form a set of equivalent bonds. Valence state is closely related to hybridization. For instance, in the case of carbon atom the V_4 valence state is a hypothetical state in which each of the sp^3 hybrid orbitals is occupied by one electron. Formation of the valence state requires energy, called promotion energy.

The oxidation state of an element in a compound is defined by the oxidation number, which is equal to the number of positive or negative elementary charges which can be attributed to the atom, provided bonds are purely ionic. In order to determine the oxidation number the following rules are followed:

1. The sum of oxidation numbers of all atoms in a neutral molecule is equal to zero and is equal to the charge of an ionic species.
2. We assign a negative charge to a more electronegative atom and a positive charge to a less electronegative atom.
3. Fluorine atoms always show oxidation number -1 and oxygen atoms -2, except in peroxides and superoxides for which the oxidation number is -1 and $-\frac{1}{2}$, respectively. In OF_2 oxygen shows oxidation number $+2$.
4. Each M atom in an M–M bond shows oxidation number 0. Thence, for instance, the oxidation number of each Hg atom in ClHg–HgCl is $+1$.

It should be noted that the effective charge on a bonded atom is less than its oxidation number, particularly for high oxidation states.

5.2 OXIDATION STATES OF s AND p BLOCK ELEMENTS

The s elements show oxidation numbers equal to the Group number, whereas the p block elements also show oxidation numbers two below the (old) Group number. For instance tin shows oxidation numbers $+4$ and $+2$ (the old Group number is IVA). The tendency to show the lower oxidation state is attributed to the inert pair effect. Since the electron configuration of the p block elements in the oxidation state two below the (old) Group number is ns^2 one can presume that removal of the two s electrons is particularly difficult. However, as we shall see below, the inert pair effect is not solely a result of high ionization energies for the s electrons. The tendency of the p block elements to display oxidation numbers two below the Group number can be generalized by stating that valence and oxidation number of these elements almost always change by two. For instance, sulphur is di-, tetra- and hexavalent and shows oxidation numbers -2, $+2$, $+4$ and $+6$, while iodine shows oxidation numbers -1, $+1$, $+3$, $+5$ and $+7$. This raises three questions:

– why Group 2 and 12 elements (except for dimeric cations of Group 12 elements) do not show the oxidation number $+1$;
– why oxidation numbers of p block elements change as a rule by two units, while the oxidation states of d elements may change by one unit (after reaching oxidation state $+2$ or $+1$, the latter in the case of Cu and Ag). A good example is manganese, which in its ions and complexes shows all oxidation numbers from $+2$ to $+7$;
– why the stability of the higher oxidation state of p block elements decreases down each Group.

To answer the first question let us examine the enthalpy of the disproportionation reaction

$$2\,Mg^+aq \rightarrow Mg^0(s) + Mg^{2+}aq \qquad (5.1)$$

For this purpose we construct the following cycle

$$
\begin{array}{ccc}
2\,Mg^+(g) & \xleftarrow{\;I_1 - I_2\;} & Mg^0(g) \;+\; Mg^{2+}(g) \\[1em]
\Big\downarrow 2\,\Delta H_{hydr}(Mg^+) & \uparrow \Delta H_{atom} & \Big\downarrow \Delta H_{hydr}(Mg^{2+}) \\[1em]
2\,Mg^+aq & \xrightarrow{\;\Delta H_{dispr}\;} & Mg(s) \;+\; Mg^{2+}aq
\end{array}
$$

in which the subscripts $_{dispr}$, $_{atom}$ and $_{hydr}$ denote disproportionation, atomization and hydration, respectively. In a thermodynamic cycle the sum of the enthalpy terms equals zero, therefore

$$\Delta H_{dispr} = -2\Delta H_{hydr}(Mg^+) - I_1 + \Delta H_{hydr}(Mg^{2+}) + I_2 - \Delta H_{atom}. \qquad (5.2)$$

Strictly ΔG°_{dispr} provides the correct measure of the tendency to disproportionate, but ΔH_{dispr}, ΔH_{hydr} and ΔH_{atom} are much more accessible quantities than their Gibbs free energy analogues. Also I is an energy, here equivalent to an enthalpy. Nonetheless ΔH_{dispr} does here provide an acceptable guide.

Table 5.1 – Factors affecting the stability of Mg^+ and Mg^{2+} ions in aqueous solution. The radii $\langle r_{3s} \rangle$ and r_i of Mg^+, and $\Delta H_{hydr}(Mg^+)$, are estimated values.

	Mg^0	Mg^+	Mg^{2+}
Configuration	$[Ne]\,3s^2$	$[Ne]\,3s^1$	$[Ne]$
$\langle r_{out} \rangle$ / pm	$172\;(\langle r_{3s} \rangle)$	$\approx 160\;(\langle r_{3s} \rangle)$	$36\;(\langle r_{2p} \rangle)$
r_i (CN 6)	–	≥ 160	72
ΔH / kJ mol^{-1}	150 (atomization)	≈ -270 (hydration)	-1931 (hydration)
I_n / kJ mol^{-1}	$738\;(n=1)$	$1451\;(n=2)$	

Substitution of the data from Table 5.1 into eq. 5.2 gives

$$\Delta H_{dispr} = 540 - 740 - 1931 + 1451 - 150 = -830 \text{ kJ mol}^{-1}.$$

This highly negative disproportionation enthalpy means that the Mg^+ ion even if formed in aqueous solution would spontaneously disproportionate into Mg^0 and M^{2+}. The values shown in Table 5.1 show that the highly negative disproportionation enthalpy results mainly from the interplay between hydration enthalpies and ionization potentials. To a crude approximation we have

$$\Delta G^{\circ}_{hydr} = -\frac{Nz^2e^2}{r}\left(1-\frac{1}{D_{H_2O}}\right) \tag{5.3}$$

where D_{H_2O} denotes the dielectric permeability of water. From eq. 5.3 the Gibbs free energy (and the enthalpy) of hydration is the more negative, the higher the charge of the ion and the smaller its radius. The charge of the Mg^+ cation is low whereas its radius is very large, because it depends on the large radius of the $3s$ orbital, $\langle r_{3s}\rangle$. Hence, the hydration enthalpy of Mg^+ is small and does not compensate the energy necessary to remove the first electron from the magnesium atom. On the other hand the higher charge and much smaller ionic radius of the Mg^{2+} cation result in a hydration enthalpy contribution negative enough to prevail over the second ionization potential, I_2. The radius of the Mg^{2+} cation is small, because it depends on the very small radius of the $2p$ shell, $\langle r_{2p}\rangle$. One should notice that the radius of the outermost subshell always decreases considerably when removal of an electron exposes a lower lying subshell. This is not the case when the first electron is detached from the Mg atom but it happens with the removal of the second electron.

The ionic model can be also used to answer the second question, i.e. to explain why the oxidation number changes by two units in the case of Group 13 and 14 elements. The ionic model can be used because these elements, like Group 1 and Group 2 elements, can exist in aqueous solutions or at least in crystal lattices as cations. Factors which affect stability of oxidation states of Group 13 and 14 elements will be now discussed using the example of thallium. Let us examine the enthalpy of disproportionation of thallium in the even oxidation state +2 which is "forbidden" in an odd Group:

$$2\,Tl^{2+}aq \rightarrow Tl^+aq + Tl^{3+}aq. \tag{5.4}$$

From a cycle similar to that for magnesium the enthalpy of disproportionation is given by

$$\Delta H_{dispr} = \Delta H_{hydr}(Tl^+) - 2\Delta H_{hydr}(Tl^{2+}) - I_2 + \Delta H_{hydr}(Tl^{3+}) + I_3 \tag{5.5}$$

Substitution of the data from Table 5.2 into eq. 5.5 gives

$$\Delta H_{dispr} = -330 + 2700 - 1970 - 4184 + 2880 = -904 \text{ kJ mol}^{-1}$$

This is a highly negative value, which means that the Tl^{2+} ion is very unstable in aqueous solution. The main reason is that the hydration enthalpy of the Tl^{2+} ion is relatively small, while that of Tl^{3+} is high enough to overcome the effect of the third ionization potential. The hydration enthalpy of Tl^{2+} is relatively small for a doubly charged cation, because removal of two electrons from the Tl atom does not expose a deeper lying shell and the ionic radius of Tl^{2+} thus depends on the large radius of the $6s$ shell. On the other hand detaching the electron from the Tl^{2+} ion uncovers the deep lying $5d$ shell, which results in a small ionic radius for the Tl^{3+} ion and a highly negative hydration enthalpy that counterweighs the ionization energy. Therefore, after attaining the +2 oxidation state thallium tends to lose the next electron. It follows from the electron configuration of Group 13 and 14 elements that the radius of a cation is determined by a deeper lying shell in the atom (either

$(n-1)d$ or ns) only when the oxidation number has its maximum value or is two less. This explains why oxidation numbers change by two units. This reasoning is valid provided that the two outer electrons in Tl^+aq and, especially, the one outer electron in Tl^{2+}aq are localized in the s orbital and not in a hybridized sp^3 orbital (see Section 5.5). However, the second option seems highly improbable because water is a rather weak ligand, unable to overcome the promotion energy necessary for hybridization. The same argument applies to Mg^+aq.

Table 5.2 – Factors affecting the stability of the Tl^+, Tl^{2+} and Tl^{3+} ions in aqueous solution. The $\langle r_{6s} \rangle$ and r_i radii and the hydration enthalpy of the Tl^{2+} ion are estimated values.

	Tl^0	Tl^+	Tl^{2+}	Tl^{3+}
Configuration	$5d^{10}6s^26p^1$	$5d^{10}\,6s^2$	$5d^{10}6s^1$	$5d^{10}$
$\langle r_{out} \rangle$ / pm	186 ($\langle r_{6p} \rangle$)	136 ($\langle r_{6s} \rangle$)	$\approx$124 ($\langle r_{6s} \rangle$)	73 ($\langle r_{5d} \rangle$)
r_i / pm (CN 6)	–	150	$\geq$ 125	88.5
ΔH_{hydr}/kJ mol^{-1}	–	–330	$\approx$ –1350	–4184
I_n / kJ mol^{-1}	590 ($n=1$)	1970 ($n=2$)	2880 ($n=3$)	–

We already know that when electrons are removed from the same subshell the radius of a cation does not decrease very much and its hydration enthalpy becomes more negative, mainly because of increased charge. In such a case the higher oxidation state is not favoured by the not particularly negative hydration enthalpy and consecutive oxidation states usually do not differ very much in stability. This is the case with d electron elements, which after the s electrons (electron) are detached, form higher oxidation states at the expense of d electrons. However, removal of consecutive d electrons (except for the last one) does not uncover an inner shell. Therefore, interaction between the cation and the ligands does not experience an abrupt increase which, if it did happen, would strongly favour the corresponding oxidation state. Factors affecting the stability of oxidation states of d block elements are discussed in more detail in Section 15.3.

The ionic model which, with some reservations, can also be applied to Group 14 elements cannot explain the stability of oxidation states of Group 15–17 elements, because these elements in positive oxidation states do not form simple cations. For these elements a model based on hybridization is more appropriate. This will now be used to explain stabilities of the oxidation states of sulphur, and also of selenium and tellurium.

In accordance with the general rule concerning oxidation states of p block elements, sulphur shows, in principle, only even oxidation states. It forms with, for example, fluorine the compounds SF_2 (a fugitive species), SF_4 and SF_6. Changes in hybridization of sulphur valence orbitals and in the number of nonbonding valence electrons (n_n) in oxidation of sulphur by fluorine, assuming that oxidation of SF_4 to SF_6 proceeds through the hypothetical SF_5 intermediate, are shown in Table 5.3.

Table 5.3 – Oxidation of SF_4 by fluorine

	SF$_4$	$\xrightarrow{+F}$	SF$_5$	$\xrightarrow{+F}$	SF$_6$
Hybridization	sp^3d		sp^3d^2		sp^3d^2
n_n	2		1		0

It can be seen from Table 5.3 that in oxidation of SF_4 to SF_5 hybridization changes from sp^3d (distorted trigonal bipyramid) to sp^3d^2 (distorted octahedron), which means that the number of d orbitals taking part in hybridization increases. This increase is necessary to place five fluorine atoms and one electron around the sulphur atom. The promotion energy required to utilize an additional d orbital is more or less compensated by the bonding energy of the fifth fluorine atom. In contrast to the first step, oxidation of SF_5 to SF_6 is not accompanied by a change in hybridization. Thus no additional promotion energy is required and the total energy of attaching the sixth fluorine atom stabilizes the system. Therefore, under conditions necessary to obtain SF_5, oxidation proceeds through to formation of the stable final product, SF_6. The instability of the MX_3 binary halides (M = S, Se, Te) can be explained by the same reasoning. The SF_4, SF_5 and SF_6 molecules are all hypervalent, and because sulphur is a light member of Group 16 the model based on hybridization employing $3d$ orbitals may be controversial (see Section 5.1). However, it has been used here because of its simplicity.

It should be noted that the hybridization model could also be used to explain stabilities of oxidation states of Group 13 and 14 elements. For instance, we can assume that gallium in the GaCl molecule, in the hypothetical $GaCl_2$ molecule, and in the $GaCl_3$ molecule shows sp, sp^2 and sp^2 hybridization, with the number of nonbonding electrons 2, 1 and 0, respectively. Using the same arguments as in the case of sulphur we immediately see that Ga(II) must be unstable. However Ga(II) can be stabilized in the dimeric complex anion $[Cl_3Ga-GaCl_3]^{2-}$ which has a metal-metal bond.

Similar considerations for Group 15 and 17 elements show that, in general, a change from a stable to an unstable oxidation state with one electron less is always accompanied by an increase in the number of orbitals participating in hybridization. One should also notice that in atoms of p block elements in uncommon oxidation states one of the hybrid orbitals is always occupied by an unpaired electron. Therefore, two such radical molecules may form a relatively stable dimer.

5.3 CHANGES IN STABILITY OF THE MAXIMUM OXIDATION STATE DOWN THE p BLOCK GROUPS

It is well known that the stability of the highest oxidation state decreases, as a rule, down each Group of p block elements and is lowest for the heaviest member. For instance Tl(III), Pb(IV) and Bi(V) are much stronger oxidants than their lighter congeners in the same oxidation state. This effect will now be discussed using the Group 13 and 14 elements as examples. Table 5.4 and Fig. 5.3 show ionization potentials, radii and enthalpies of hydration for Group 13 elements.

Table 5.4 – Factors affecting the stability of the +1 and +3
oxidation states in Group 13.

	Al	Ga	In	Tl
$I_1/\text{kJ mol}^{-1}$	577	579	558	589
$I_2 + I_3 /\text{kJ mol}^{-1}$	4561	4942	4525	4849
$\langle r_{ns}\rangle/\text{pm}$	137.2	129.5	144.1	136.4
$r_i\,(1+)/\text{pm}$	–	–	~120	150
$r_i\,(3+)/\text{pm}$	53.5	62.0	80.0	88.5
$\Delta H^\circ_\text{hydr}(M^+)/\text{kJ mol}^{-1}$	$\approx-400^a$	$\approx-320^a$	-296	-310
$\Delta H^\circ_\text{hydr}(M^{3+})/\text{kJ mol}^{-1}$	-4661	-4685	-4108	-4184

^{a} Estimated values.

It is evident that oxidation of M^+ to M^{3+} is facilitated by a low value for the sum of the energies $I_2 + I_3$. Therefore, according to the data in Table 5.4 and Fig. 5.3 the stability of the 3+ oxidation state with respect to the 1+ oxidation state should vary in the following order:

$$In \approx Al > Tl \approx Ga\,.$$

On the other hand the more negative the hydration enthalpy of the M^{n+} ion the more stable will be the respective oxidation state in aqueous solution. The hydration enthalpies of the M^+ ions are small and very similar (Table 5.4). In contrast, the hydration enthalpies of the M^{3+} ions are large, with their respective values differing markedly between Al, Ga and In, Tl and suggesting the following stability order for the 3+ oxidation state:

$$Ga \approx Al > Tl \approx In\,.$$

The order of experimentally observed relative stabilities of the 3+ oxidation state with respect to the 1+ oxidation state in aqueous solution is

$$Al > Ga > In \gg Tl$$

which represents a compromise between changes in the ionization energy sum and in $\Delta H^\circ_\text{hydr}$ for the M^{3+} ions. The much lower stability of thallium than indium in oxidation state +3 results from similar $\Delta H^\circ_\text{hydr}$ but markedly greater I_2+I_3. This is, in turn, caused by relativistic effects and the presence of the filled $4f$ shell. These two factors strongly stabilize the $6s$ electrons in thallium.

Similar considerations for Group 14 elements (see Fig. 5.4) show that due to the sum of the ionization potentials I_3 and I_4, the stability of the +4 oxidation state with

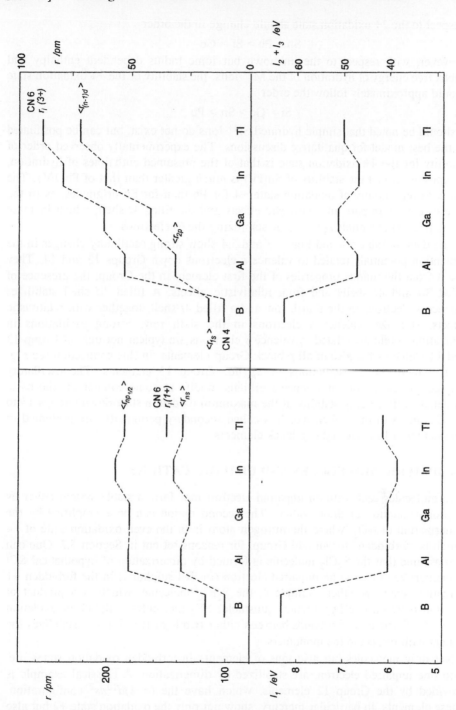

Fig. 5.3 Changes of radii and ionization energies in Group 13 elements.

respect to the 2+ oxidation state should change in the order

$$Sn > Pb > Si > Ge$$

However, with respect to the unknown but ionic radius dependent enthalpy and Gibbs free energy of hydration of the M^{4+} ions, the stability of the +4 oxidation state should approximately follow the order

$$Si > Ge > Sn > Pb .$$

It should be noted that simple hydrated Si^{4+} ions do not exist, but can be postulated as the best model for qualitative discussions. The experimentally observed order of stability for the 4+ oxidation state is that of the presumed enthalpies of hydration, except that in fact the stability of Sn(IV) is *much* greater than that of Pb(IV). The much lower stability of oxidation state +4 for Pb than for Sn originates, as in the case of the Tl, In pair, in relativistic effects and the filled $4f$ shell, which increase the ionization potentials I_3 and I_4 by stabilizing the $6s$ electrons.

The data in Table 5.4 and Figs. 5.3 and 5.4 show strong oscillatory changes in the ionization potentials related to valence s electrons down Groups 13 and 14. They result from the unique properties of the first element in the Group, the presence of filled $3d$ and $4f$ shells and from relativistic effects. A filled $3d$ shell stabilizes valence s electrons in the fourth row and a filled $4f$ shell, together with relativistic effects, stabilizes valence s electrons in the sixth row. Strong oscillations in ionization potentials related to valence s electrons are typical not only of Group 13 and 14 elements but also of all p block Group elements. In this connection see Fig. 11.1, which shows oscillations in ε_{ns} for Group 15 elements. The oscillatory changes in the energy of valence s orbitals modify in some compounds the basic tendency of decreasing stability of the maximum oxidation state down Groups 15 to 17. Examples of this effect, which is called secondary periodicity, are presented in Chapters 9 to 13 dedicated to p block elements.

5.4 DIMERIC MOLECULES AND DIMERIC CATIONS

Two molecules each with an unpaired electron may form a stable system either by disproportionation or dimerization. The second option can be exemplified by the dimerization of NO_2, where the nitrogen atom is in the even oxidation state of +4 which is "forbidden" for an odd Group, for reasons set out in Section 5.2. One can also assume that the S_2Cl_2 molecule is formed by dimerization of hypothetical SCl monomers each having an unpaired electron (in SCl sulphur is in the forbidden +1 oxidation state). Another example is the S_2F_{10} molecule, which is a product of reaction between two hypothetical (unstable) SF_5 molecules (sulphur in oxidation state +5). Formation of σ bonds between either two N or two S atoms stabilizes the dimers with respect to the monomers.

Not only molecules but also ions of elements in forbidden oxidation states, i.e. with one unpaired electron, are stabilized by dimerization. A classical example is provided by the Group 12 elements, which have the $(n-1)d^{10}ns^2$ configuration. These elements, in particular mercury, show not only the oxidation state +2 but also +1, which is forbidden for an even group. However, the 1+ cations of Group 12

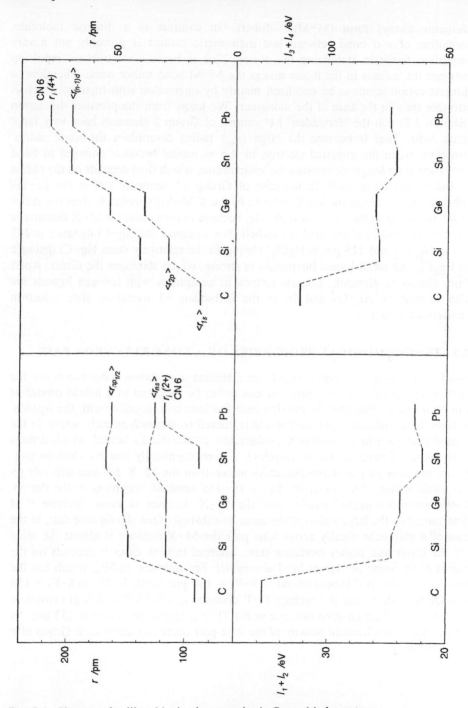

Fig. 5.4 Changes of radii and ionization energies in Group 14 elements.

elements always form $(M-M)^{2+}$ dimers. In contrast to a dimeric molecule, formation of a σ bond between two monomeric cations is probably not a very important factor in stabilizing the dimer. This is because electrostatic repulsion between the cations in the dimer makes the M−M bond rather weak. Therefore, a dimeric cation seems to be stabilized mainly by interaction with ligands, which is stronger than in the case of the monomer. We know from the previous discussion (Section 5.2) that the "forbidden" M^+ cations of Group 2 elements have very large ionic radii. That is because the large $\langle r_{ns} \rangle$ radius determines the ionic radius. However, when the unpaired electron in the ns orbital becomes engaged in bond formation it no longer determines the ionic radius, which then depends on the radius of the deeper lying shell. In the case of Group 12 elements this is the $(n-1)d$ subshell. As the result the M−X distance in the X[M−M]X molecule does not differ much from that in the X−M−X molecule, because in both cases the M−X distance is fixed by the radius of the $(n-1)d$ subshell. For instance, the Hg−Cl distance is 243 pm in Hg_2Cl_2 and 225 pm in $HgCl_2$. Owing to the relatively short Hg−Cl distance in Hg_2Cl_2 the metal-ligand interaction is strong, which stabilizes the dimer. Apart from Group 12 elements, dimeric cations in complexes with halogen ligands are also formed by Al, Ga and In in the forbidden +2 oxidation state (electron configuration ns^1).

5.5 STEREOCHEMICAL PROPERTIES OF A LONE ELECTRON PAIR

The lone electron pair appears in p block elements in oxidation states two below the highest group valence. The lone pair can either be localized in a hybrid orbital or can retain its s^2 character. In the first case the lone pair, together with the ligands, defines the coordination polyhedron and is, therefore, stereochemically active. In the second case the pair remains in the spherically symmetrical s orbital, which defines only the M−X distance and is, therefore, a stereochemically inactive electron pair. When the lone pair is stereochemically active then the M−X distance depends on the small radius of the inner $(n-1)p$ or $(n-1)d$ subshell, according to the Period. For a stereochemically inactive pair the M−X distance is large, because it is determined by the large radius of the outer ns subshell. One should note that, in the case of a stereochemically active lone pair the M−X distance is almost the same for the lower and higher oxidation state, because in both cases it depends on the radius of the same $(n-1)p$ or $(n-1)d$ subshell. For instance, in SF_4, which has the lone pair in the sp^3d hybrid orbital, $d(S-F) = 160$ pm, while in SF_6 $d(S-F) = 156$ pm. On the other hand the average Tl−F distance in solid TlF is 279 pm (inactive pair) whereas the analogous distance in $Na[Tl^{III}F_4]$ (active pair) is only 233 pm. As a rule, the stereochemical activity of the lone pair decreases down each Group of p block elements.

6

Catenation and formation of condensed phases

6.1 CATENATION

Catenation can be defined as self-linking of atoms of an element to form chains and rings. This definition can be extended to include formation of layers (two-dimensional catenation) and space lattices (three-dimensional catenation). One- and two-dimensional catenation, together with weaker interactions between the chains, rings or layers leads to formation of solid phases in which not all bonds between atoms are equivalent. The necessary condition for catenation is a valence of at least two, with the notable exception of Group 17 elements (see p. 58) and formation of strong bonds between atoms of the same element. Since the bond energy between two atoms of an element decreases with increasing radius down each p block Group, Table 6.1, catenation shows the same tendency. When catenation requires the presence of an "auxiliary" X atom (an atom which does not enter the framework, e.g. the H atom in hydrocarbons) high A−X energy also favours catenation.

Table 6.1 − Radii of atoms and single bond energies of Group 14 elements.

Element	C	Si	Ge	Sn
Radius of the atom, R / pm	92	147	152	168
A−A single bond energy / kJ mol^{-1}	≈ 350	≈ 230	≈ 190	≈ 130

Catenation in the Groups

Groups 1 and 11.

Because the elements are monovalent, catenation based on 2c-2e bonds is precluded.

Groups 2, 12 and 13.

The number of valence electrons makes one-dimensional catenation theoretically possible with sp or sp^2 hybridization (for sp^2 with an auxiliary atom). However, except for boron the elements of these Groups prefer formation of metallic phases. Formation of the B_{12} icosahedron which is the basis for all allotropic forms of boron can be considered as two-dimensional catenation. However, a better description is in terms of cluster formation. Formation of boranes (boron hydrides) is another example of the tendency of boron atoms for self-linking. Among Group 12 elements mercury catenates relatively easily to form linear polymercury cations Hg_n^{2+} with n = 2 to 4. Oxidation of mercury by arsenic pentafluoride at low $AsF_5:Hg$ ratios gives solids of the general formula $Hg_n(AsF_6)_2$ containing infinite chains of mercury atoms separated by columns of AsF_6^- anions.

Group 14.

Carbon is certainly the best catenator among all elements. The various catena-forms of carbon depend on the hybridization of the carbon atoms. The different modes of catenation described below are formal descriptions, not real courses of reactions.

1. *sp*-Hybridization

sp-Hybridization is shown by carbon atoms in ethyne (acetylene), $HC \equiv CH$, which has one σ and two $p\pi-p\pi$ bonds. The hydrogen atoms in ethyne can by replaced by successive $-C \equiv C-$ alkyne groups, which then leads to formation of an "infinite" chain $[-C \equiv C-]_n$. Due to van der Waals interactions the alkyne chains stack together and form the allotropic modification of carbon called chaoite or carbyne carbon. Chaoite is stable above $2600\,°C$, metastable at room temperature.

2. *sp*²-Hybridization

In this hybridization each of the three sp^2 hybrid orbitals is occupied by one electron. The fourth electron occupies the p orbital perpendicular to the triangle. Two sp^2 hybridized carbon atoms may form the $>C=C<$ structural unit which contains a double bond. Attachment of four hydrogen atoms results in formation of ethene, $H_2C=CH_2$. If only two hydrogen atoms are attached to the $>C=C<$ structural unit then the three $-HC=CH-$ elements may self-link forming the benzene ring. An "infinite" number of sp^2-hybridized carbon atoms forms by self-linking a planar layer which is an example of two-dimensional catenation. The unpaired electrons in the p orbitals perpendicular to the plane form delocalized $p\pi-p\pi$ bonds. Weak van der Waals interactions between the layers lead to formation of the allotropic form of carbon called graphite. Carbon atoms in the sp^2 state form an incredible variety of recently discovered structures called fullerenes and nanostructures which resemble rolled graphite sheets. However, in contrast to graphite these spherical (ellipsoidal) or tubular structures consist of both hexagons and pentagons. Fullerenes are usually described as clusters.

3. *sp*³-Hybridization

In this form of hybridization each of the four carbon hybrid orbitals is occupied by one electron. Self-linking of carbon atoms by σ-bonds leads to formation of the diamond lattice and is an example of three-dimensional catenation. Silicon and germanium catenate in the same way. Alternatively, each carbon atom in the tetravalent state can attach two hydrogen atoms and form the $-CH_2-$ structural unit. A number of such structural units may self-link and form aliphatic hydrocarbons $CH_3-(CH_2)_n-CH_3$ or alicyclic hydro-carbons e.g. *cyclo*-C_6H_{12}, cyclohexane. These are classical examples of mono-dimensional catenation. In the same way hydrides of silicone (silanes) and of germanium (germanes) are formed. Because of the decreasing strength of the respective A–A bonds silicon forms Si_6H_{14}, germanium Ge_9H_{20} (which is rather unstable), tin Sn_2H_6, but lead only PbH_4.

Group 15.

Catenation is possible in the sp^3 valence state. This requires participation of hydrogen as the auxiliary atom and leads to formation of a chain, with a lone electron pair on each A atom :

$$H_2A-(AH)_n-AH_2 .$$

The tendency to catenate is very weak. For nitrogen $n = 0$ (hydrazine), and for phosphorus the maximum value of n is 1 (triphosphine). However, making use of only p orbitals phosphorus readily catenates to form rings, double layers or long chains as shown schematically below. In these structures there is one lone electron pair (the s^2 pair) on each P atom.

Group 16.

Oxygen forms only the ozone molecule, O_3, and the recently discovered unstable O_4. As we know from Section 3.2, oxygen atoms, because of their small size, readily form $p\pi-p\pi$ bonds with each other. It appears that formation of small O_3 molecules, each with a delocalized $p\pi-p\pi$ bond in addition to the two σ bonds, is energetically more efficient than formation of a long chain, see Section 12.1. The much larger sulphur atoms are reluctant to form $p\pi-p\pi$ bonds, hence they catenate almost as eagerly as carbon atoms. From the four hybridized sp^3 orbitals in the sulphur atom two are occupied by single electrons and two by pairs. Therefore, sulphur atoms can self-link by σ bonds to form stable chains without auxiliary atoms. Self-linking leads to formation of two basic forms:

1. *Cyclo-*S_n where n may vary from 6 to 20. The most important cyclic form is S_8, which is basis for many allotropic forms of sulphur, among them the most stable allotrope the orthorombic α form and the monoclinic β and γ forms.

2. *Catena-*S_∞. These are helical chains which by sticking together form the so-called fibrous sulphur and are present in the plastic, polymeric and other ill-defined forms of sulphur.

Polysulphides M_2S_n ($n = 2-6$) and polythionic acids $HO_3S-S_n-SO_3H$ ($n = 1-4$) are other examples of the ability of sulphur to catenate. Selenium and tellurium also catenate but less readily, because the greater radii of the atoms result in reduced bond energy.

Sulphur and nitrogen atoms linking alternately form $(SN)_x$ chains. In spite of the presence of one unpaired electron in the third p orbital, the nitrogen atom does not attach hydrogen as the auxiliary atom to form the expected $(SNH)_n$ chain. Instead, the p orbital on the N atom and the (empty) d orbital on the S atom form a $p\pi-d\pi$ bond. The $p\pi-d\pi$ bonds in the $(SN)_n$ chain conjugate and the unpaired electrons become delocalized. In terms of the band model one can say that the p orbitals of nitrogen and d orbitals of sulphur form the conduction band. The compound $(SN)_n$ is a one-dimensional metal with electrical conductivity along the chain similar to that of mercury. Sulphur-bonded nitrogen can attach a hydrogen atom but then, instead of a chain, eight-membered $S_{8-n}(NH)_n$ rings ($n = 1 - 4$) are formed. Since with respect to the number of outer electrons the NH group is isoelectronic with the S atom, the sulphur-imido compounds can be considered to result from replacing the S atom in the S_8 ring by the NH group.

Group 17.

According to the simple theory of catenation halogen atoms should not self-link because, like alkali metals, they have only one unpaired electron. Formation of brown-coloured polyiodide anions (I_3^-, I_5^-, I_7^- and I_9^-) is possible due to formation of multicentred bonds, instead of the usual two-centre, two-electron (2c–2e) bonds. In the simplest and most stable polyiodide anion, I_3^-, the central iodine atom is surrounded by three lone pairs of electrons in equatorial positions, with the two outer iodine atoms occupying the axial positions of the trigonal bipyramid. The I–I–I axial bond is a 3c-4e bond and holds the two extra electrons beyond the octet in the non-bonding orbital, which belongs only to the two apical iodine atoms. With respect to the arrangement of electrons around the central atom, the I_3^- ion is identical with the PF_5 molecule, and is a hypervalent species (see Section 5.1). Formation of higher polyiodide anions can be formally described in terms of consecutive additions of I_2 molecules to the I_3^- anion. Structures for some polyiodides are shown in Fig. 6.1. Interatomic distances in these and other polynuclear iodine species are discussed towards the end of Chapter 13, where they are linked to heteronuclear polyhalides and other related aspects of halogen chemistry.

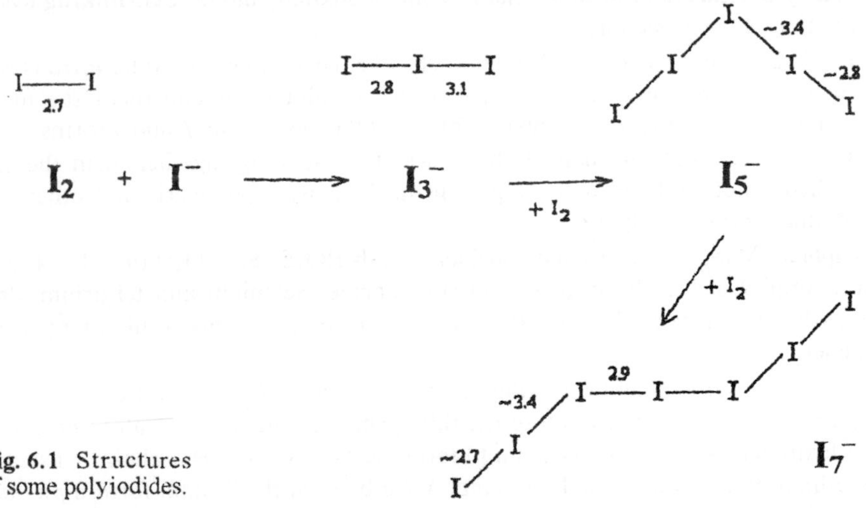

Fig. 6.1 Structures of some polyiodides.

Elements of the d *block.*

As a rule atoms of the transition elements do not catenate, but show strong tendencies to self-linking with formation of clusters.

6.2 FORMATION OF CLUSTERS

Formation of clusters bears some resemblance to catenation. In both forms of aggregation atoms of an element self-link by (usually) σ bonds. However, catenation leads to formation of chains and rings, while in clusters atoms usually occupy corners of a polyhedron. Bonding in a cluster is very different from that in a coordination compound. In a coordination compound ligands which occupy the corners of a polyhedron are bonded to the central atom but not to each other. On the other hand (with a few exceptions – see Section 10.3) there is no central atom in a cluster, and "ligands" occupying the corners of a polyhedron (usually atoms of the same element) are linked by chemical bonds. The necessary condition for cluster formation is high bond energy which, in turn, requires low oxidation state of the element (low charges on the atoms in the framework) in order to diminish electrostatic repulsion. The type of coordination polyhedron depends on the number of valence orbitals and electrons in the atoms which form the cluster framework.

Both Main Group and transition elements form clusters. The atoms that form the framework of a cluster may or may not have additional atoms or groups attached. Clusters of p block elements usually do not require additional atoms or groups to complete the valence shell of the framework atoms. Examples of p-block clusters are the P_4 molecule and the anionic clusters $[Pb_5]^{2-}$, $[Sn_9]^{4-}$ and $[As_7]^{3-}$. Anionic clusters are formed when a moderately electronegative element (P, As, Sb, Sn or Pb) reacts with a highly electropositive metal from Group 1 or 2. In such a reaction valence electrons are almost completely transferred from the very electropositive metal atoms to the atoms of the p block elements which then form negatively charged polyhedra. In contrast to p block elements, clusters of d block elements generally contain additional atoms or groups. In the case of early d block elements, which are electron-poor, the attached species are electron donors such as the Br^- anion. On the other hand clusters formed by the end members of each d series attach electron acceptors which remove some electrons from the metal. An example of the first type is the octahedral $[W_6Br_8]^{4+}$ cluster and of the second the tetrahedral $[Co(CO)_3]_4$ cluster.

The simplest examples of bonding between like atoms are diatomic molecules of alkali metals and halogens, the $ClHg-HgCl$ molecule and the $[Cl_4Re-ReCl_4]^{2-}$ anion. Examples of triangular species are the $[Re_3Cl_{12}]^{3-}$ anion and the $Os_3(CO)_{12}$ molecule.

Typical representatives of proper clusters (atoms in the corners of a polyhedron) are the tetrahedral P_4, As_4 and Sb_4 molecules. In these molecules each atom is linked to the other three atoms by σ bonds formed by p orbitals, while its lone electron pair remains in the s orbital. Since the p orbitals are perpendicular to one another whereas the angle in an equilateral triangle is $60°$ the bonds in the P_4 molecules are strained, so that the bond energy is low. That is the reason why the reactivity of white phosphorus, which is formed by stacking the P_4 molecules, is so high. Carbon forms clusters which contain many atoms. The best known is the C_{60} cluster – a polyhedron which has 12 pentagonal and 20 hexagonal faces. Clusters, just as the catena-forms, interact by van der Waals forces and form weakly bonded solid phases.

In contrast to coordination compounds the cube is a polyhedron frequently found among clusters. A classical example is the hydrocarbon called cubane, C_8H_8 or $(CH)_8$. The icosahedron formed by 12 B atoms is the basic unit of all allotropic forms of elementary boron. Also borane anions (closo-boranes) $B_nH_n^{2-}$ and the corresponding carboranes $B_{n-2}C_2H_n$ ($n = 5-12$) are typical clusters. $B_6H_6^{2-}$ and $B_4C_2H_6$ have octahedral frameworks, while in $B_{12}H_{12}^{2-}$ and $B_{10}C_2H_{12}$ the boron (and carbon) atoms occupy the corners of an icosahedron.

Clusters can be divided into two groups: simple and mixed clusters. In simple clusters there is only one kind of atom in the framework. In mixed clusters the corners of a polyhedron are either occupied by two kinds of atoms or the framework consists of two superposed polyhedra each containing atoms of one kind only. For example, in carboranes boron and carbon atoms occupy vertices of the same polyhedron, whereas $N_4(CH_2)_6$, $As_4(NCH_3)_6$ and adamantane, $C_{10}H_{16}$ or $(CH)_4(CH_2)_6$, are examples of mixed clusters containing a tetrahedron and an octahedron. Another example of this type of cluster is S_4N_4, in which S atoms occupy the corners of a tetrahedron and N atoms the corners of a square.

Fe_4S_4 clusters form an integral part of a number of electron transfer proteins, while the mixed cubic cluster containing alternating iron and sulphur atoms with one of the Fe atoms replaced by an Mo atom plays a central role in fixing atmospheric nitrogen. This cluster can be found in the enzyme nitrogenase produced by some bacteria. In the fixation process the N_2 molecule becomes attached to the Mo atom, which decreases the strength of the triple bond in the N_2 molecule and facilitates its reduction.

6.3 FORMATION OF CONDENSED PHASES

At sufficiently high temperature each element forms atoms in the gas phase. When the temperature is decreased to its room value the atomic gas may choose between a number of options.

1. It can remain as an atomic gas, as the noble gases do. The atoms of some other elements may self-link and either stay as a molecular gas (H_2, N_2, O_2, F_2, Cl_2, O_3), or condense to form phases containing small molecules as e.g. P_4 in white phosphorus, S_8 molecules in orthorhombic sulphur, Br_2 and I_2 molecules in liquid bromine and solid iodine.
2. Solid phases containing macromolecules can be formed. Examples are carbyne carbon and fibrous sulphur. One can also include in this category solid structures consisting of layers as e.g. black phosphorus.
3. Formation of solid phases with equivalent, in principle, positions of atoms and bonds may take place. Examples are metallic phases and the covalent phase of the diamond type exhibited also by Si, Ge and α-Sn. In the case of the diamond-type lattice each atom must provide four orbitals and four electrons, while in the case of a metallic phase the number of electrons must be, as will be shown in Section 6.4, less than twice the number of orbitals.

All these processes compete with each other, and the outcome depends on which system is thermodynamically more stable. This, in turn, depends on the number of valence electrons and available orbitals, on the orbital energy and on the radius of the atom.

The outcome of the competition between formation of diatomic molecules and formation of a condensed phase of the kind in points 2 and 3 above can be shown by the example of the elements from lithium to fluorine. For this purpose one has to compare half of the enthalpy of dissociation of the M_2 molecule, $\frac{1}{2}\Delta H_{diss}$, with the enthalpy of atomization ΔH_{atom}. We take into account half of the dissociation enthalpy in order to have the same stoichiometry in the right-hand part of each of the two equations shown below.

$$M(g) \rightarrow \frac{1}{2}M_2(g) \qquad -\frac{1}{2}\Delta H_{diss} \qquad (6.1)$$

$$M(g) \rightarrow M(s) \qquad -\Delta H_{atom} \qquad (6.2)$$

From equations 6.1 and 6.2 a condensed phase is more stable than a diatomic molecule if $\Delta H_{atom} > \frac{1}{2}\Delta H_{diss}$. It follows from the data in Table 6.2 that Li, Be, B and C prefer formation of a solid phase, either metallic or covalently bonded. The small enthalpy gain in condensing the M_2 molecules does not markedly affect the outcome. For nitrogen, oxygen and fluorine formation of a covalent phase of the diamond type is forbidden, because the number of valence electrons is too great. However nitrogen could, in principle, form a covalent solid phase similar to that of black phosphorus. On the other hand formation of a metallic phase is disfavoured for reasons to be explained below. Therefore, the elements from nitrogen to fluorine prefer formation of diatomic molecules. In the case of the transition elements the metallic phase always has the upper hand over diatomic molecules.

Table 6.2 – Enthalpies of dissociation of dimeric molecules and enthalpies of atomization of some *s* and *p* elements

Element	Li	Be	B	C	N	O	F
$\frac{1}{2}\Delta H_{diss}$ / kJ mol^{-1}	54	≈ 0	145	315	473	249	79
ΔH_{atom} / kJ mol^{-1}	159	324	563	717	$\approx 300^*$	–	–

* Estimated for the structure adopted by black phosphorus

6.4 CONDITIONS FOR FORMATION OF A METALLIC PHASE

To form a σ-bonded diatomic molecule each atom must provide one orbital. Two atomic orbitals, one on each atom, combine to give two molecular orbitals – one bonding and one antibonding, Fig. 6.2(a). Each of the two molecular orbitals can accommodate two electrons. When each of the atoms provides two electrons then bonding and antibonding orbitals are filled and the bond order is zero. This is the reason why Group 2 and 12 elements do not form homonuclear diatomic molecules. When the number of periodically arranged atoms is increased, more orbitals are added and a band containing discrete energy levels with small spacing is formed,

Fig. 6.2(b). The lower part of the band corresponds with most bonding and the upper part with most antibonding orbitals.

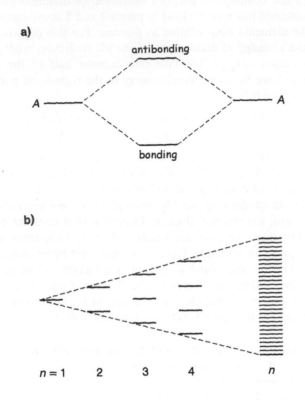

Fig. 6.2 (a) Formation of bonding and antibonding molecular orbitals from two equivalent atomic orbitals. (b) Formation of a band by combination of *n* atomic orbitals on periodically arranged atoms.

Provided the energy difference between orbitals from different subshells in the atoms is not too big, the bands arising from different orbitals may overlap and form a single energy band as e.g. the *s/p* band. The overlapping of *s* and *p* bands can be explained in terms of hybridization of orbitals in atoms in the following way. Hybrid orbitals formed by *s* and *p* orbitals on the same atom, when engaged in bond formation between two atoms, form bonding and antibonding hybrid orbitals. With increasing number of atoms bonding and antibonding *sp* orbitals form a common band. However, when the energy difference between the *sp* bonding and the *sp* antibonding orbital in the molecule is high, then two separate energy bands are formed.

If each of the atoms forming a solid phase contributes k orbitals to the band which contains both bonding and antibonding orbitals (which is not the case with semiconductors and non-metals), then the total capacity of the band with respect to electrons is $2kN$, where N is the number of atoms. On the other hand if each atom provides z electrons the total number of electrons is zN. Since at least some of the antibonding orbitals in the band should remain unoccupied in order to have net bonding energy the following relation must hold:

$$zN < 2kN \qquad (6.3)$$

or

$$z < 2k. \qquad (6.4)$$

Thus, the formation of metallic bonding, like formation of σ-bonded M_2 molecules, requires the number of electrons contributed by the atom to the band to be less than twice the number of contributed orbitals. Bonding is metallic, because incomplete filling of the band makes it possible for the electrons to move freely under an applied electric field. A general condition for formation of a metallic bond is that the relevant ionization potentials are low. The qualitative explanation is that when metallic bonds are formed atoms can be considered to be partially ionized, since electrons in metals are delocalized and do not belong to a particular atom. Metallic bonding is a form of electron-deficient bonding. For instance, in the case of metallic potassium, which forms the body-centered cubic lattice, there is only ¼ of an electron for each bond. The strength of the metallic bond, conveniently measured by the enthalpy of atomization, increases with Z in rows of d elements until the band formed by s and d orbitals is half-filled and then decreases, Fig. 6.3. The reason for perturbations in the $3d$ and $4d$ series is discussed in Section 15.2.

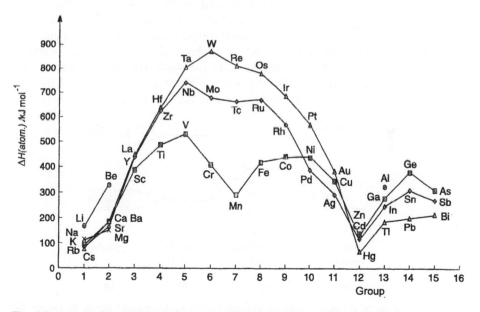

Fig. 6.3 Atomization enthalpies of s, p and d block metals (also Ge, As and Sb).

6.5 THE CHARACTER OF THE CONDENSED PHASE AND THE POSITION OF THE ELEMENT IN THE PERIODIC TABLE

Group 1, 2 and 12.

Elements in these Groups have fewer valence electrons than the number necessary for formation of covalent solid phases. Therefore, they form metallic phases where bonding is electron-deficient. Because of the small number of bonding electrons bond energy is low, in particular for the alkali metals which have very low melting and boiling temperatures. Were the energy band formed only by *s* orbitals then Group 2 and 12 elements could not form metallic bonding, just as they do not form diatomic molecules. This is because for $k = 1$ and $z = 2$ all the bonding and antibonding orbitals in the band would be filled. However, in Group 2 and 12 elements the *s* and *p* bands overlap, so that $k = 4$, $z = 2$ and metallic bonding becomes possible.

Group 13.

Atoms of Group 13 elements still have fewer valence electrons than required for formation of covalently bonded solid phases with all bonds equivalent. Therefore the elements from Al to Tl form metallic phases. Boron, because of its high ionization potentials, forms covalent solid phases based on the B_{12} cluster, see Section 9.1.

Group 14.

With respect to the number of valence orbitals and electrons, elements of this Group meet the necessary conditions for formation of both covalent and metallic solid phases. C, Si, Ge and α–Sn form the diamond-type lattice with four two-electron bonds. Pb, Sn in its β-allotrope, Si and Ge under high pressure adopt metallic phases. In diamond or in a diamond-type solid the band structure consists of two bands which are formed in the following way. We assume that one *s* and three *p* orbitals on each atom, Fig. 6.4(a), hybridize to give four sp^3 orbitals, Fig. 6.4(b).

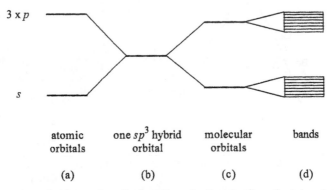

atomic orbitals	one sp^3 hybrid orbital	molecular orbitals	bands
(a)	(b)	(c)	(d)

Fig. 6.4 Formation of valence (bonding) and conduction (antibonding) bands by Group 14 elements. (a) Atomic orbitals. (b) Formation of an sp^3 hybrid orbital. (c) Two sp^3 hybrid orbitals on two adjacent atoms form a bonding and an antibonding orbital. (d) With increasing number of atoms the sp^3 orbitals broaden into filled valence and empty conduction bands.

Two sp^3 hybidized atoms combine to give two molecular orbitals, one bonding and one antibonding, Fig.6.4(c). With increasing number of periodically arranged atoms two bands are formed, the lower containing the sp^3 bonding orbitals and the upper containing the antibonding sp^3 orbitals, Fig. 6.4(d). Since the lower (valence) band contains only bonding orbitals its capacity for electrons is $\frac{1}{2}(2kN)$, and since each carbon atoms contributes 4 orbitals and 4 electrons the band is filled. The upper (conduction) band formed by antibonding sp^3 orbitals is separated from the lower by an energy gap and is empty. Because of their small radius, bonding between carbon atoms is strong which results in large difference between bonding and antibonding sp^3 orbitals. This is the reason why the corresponding bands in diamond are separated by a large energy gap. With increasing size of the atom and decreasing strength of the bond down the Group the energy difference between bonding and antibonding orbitals decreases, which also makes the energy gap between the bands decrease (see Table 10.1). Finally the two bands overlap at β-Sn and at Pb, so that electrons can move freely and the two elements become metals. One should notice that the change from the diamond-type structure to a metallic structure is accompanied by an increase in coordination number from four in the diamond-type lattice to twelve in metallic lead. That suggests that $6d$ orbitals in lead (and in polonium) participate in hybridization and the band in the metal may have $s/p/d$ character.

Groups 15-17.

Elements of these Groups have too many electrons to form non-metallic condensed phases with all bonds of the same character. Instead they form:

1. molecular phases as e.g. white phosphorus, orthorhombic sulphur or halogens – all these phases contain discrete, covalently bonded molecules held together by van der Waals forces;

2. phases containing long chains e.g. red phosphorus and fibrous sulphur;

3. layered quasi-metallic phases as e.g. the black allotrope of phosphorus, arsenic, antimony and bismuth. Atoms of these elements form double layers in which each atom is bonded to three neighbors by its p orbitals. Stacking of these double layers leads to flaky structure like that of graphite. Interaction between the layers brings about metallic conduction.

The band structure of those elements of Groups 15 to 17 which form solid phases consists of three bands: a low-lying filled bonding band, a filled nonbonding band and a high-lying empty antibonding band. In black phosphorus each P atom donates three orbitals and three electrons to the bonding band and one orbital and one electron pair to the nonbonding band, leaving the antibonding band empty. In catena-sulphur each S atom donates two orbitals and two electrons to the bonding and two orbitals and two electron pairs to the nonbonding band. Again the antibonding band remains empty. As in Group 14, and for the same reasons, the gaps between the bands decrease down Groups 16 and 17 and disappear for the heaviest member. Since in the common band which is then formed not all orbitals are occupied, electrons can move freely under the applied electrical field and the solid exhibits metallic properties. This is why polonium is a genuine metal, astatine would be a metal if it could form a condensed phase, and iodine becomes a metal under high pressure (see Section 13.2). It should be noted that when sufficiently high pressure is applied to a solid the energy gaps between the bonding, nonbonding, and antibonding bands drop to zero and the solid acquires metallic character.

Elements of the d *and* f *blocks.*

All *d*- and *f*-block elements form only metallic phases. Transition elements show an *s*/*p* band and a narrow *d* band whose energy spread lies within the *s*/*p* band (Section 15.2). Because the number of bonding electrons in transition elements is on the whole large, atomization energies and melting point temperatures are high, particularly near the middle of the series (Figs. 6.2 and 15.1). In the case of *f*-block elements the band is also formed by *s*, *p*, and *d* orbitals. In addition the *f* orbitals form a very narrow *f* band. In spite of its being very narrow, which implies significant localization of electrons, the *f* band affects properties of early members of each series, in particular the properties of early actinide metals (see Section 18.2).

6.6 ALLOTROPY

The propensity of an element to form solid phases with different structures is called allotropy. Differences in structure and bonding type between allotropic forms can either be very small as e.g. between cubic close-packed and hexagonal close-packed structures or quite big as between diamond and graphite, where the two allotropes show different types of bonding. The enthalpy for a change from one allotropic form to another is usually low. For instance ΔG° for the transformation graphite $\rightarrow$ diamond is only 2.9 kJ mol^{-1}. Each allotropic form is stable in a certain range of temperature and pressure. However, if the activation energy for transformation is high the allotrope unstable under normal conditions may become metastable. Allotropy is not very common among metals, where it is mainly shown by some 3*d* elements, light lanthanides, and actinides from Th to Cm. Differences between allotropic forms of metals are rather small. On the other hand formation of allotropic modifications is common among Group 14−16 elements. The allotropes may differ either in bonding (covalent *vs.* metallic), which is the case with α- and β-Sn (grey and white tin), or in the formation of different catena-forms, which is typical for allotropes of phosphorus and sulphur. In both cases the differences are large.

PART 2 : *Chemical properties*

7

Group 1. Hydrogen and the alkali metals

Table 7.1 — Fundamental properties

	H	Li	Na	K	Rb	Cs	Fr
$R = \langle r_{ns}\rangle / \text{pm}$	79	205	222	276	293	322	313
I_1 / eV	13.60	5.32	5.14	4.34	4.18	3.89	4.15
A / eV	0.75	0.62	0.55	0.50	0.49	0.47	0.45
χ	2.20	0.98	0.93	0.82	0.82	0.79	0.7
mp / °C	−259	180.5	97.8	63.6	39.0	28.4	
r_{met} / pm		152	186	227	248	265	
$\Delta H_{diss} / \text{kJ mol}^{-1}$ [a]	436	108	73	50	47	44	
r_i / pm (CN 6)	b	76	102	138	152	167	~180
$-\Delta H_{hydr} / \text{kJ mol}^{-1}$ [c]	1094	522	407	324	299	274	
$E°(+1/0) / \text{V}$	0.0	−3.04	−2.71	−2.92	−2.92	−2.92	~ −3.1

[a] Enthalpy of the reaction $M_2 \rightarrow 2M$. [b] $r_i(H^-) = 154$ pm in CsH. [c] Values based on $\Delta H_{hydr} (H^+) = -1094$ kJ mol^{-1}.

Table 7.1 shows that with respect to all fundamental properties of the atom, hydrogen differs considerably from the rest of the Group 1 elements. Also chemically hydrogen is very different from its heavier congeners. The remarkable chemical properties of hydrogen are the result of its high ionization energy, very small radius, and an electronegativity much higher than that of the alkali metals. The highly negative 1s orbital energy and small orbital radius result from the fact that 1s is the first atomic shell and that in the hydrogen atom there is no interelectronic repulsion. In contrast, alkali metals show strong repulsion between ns and deeper lying $(n-1)s$ orbitals which have the same symmetry. Thus, big differences between hydrogen and the alkali metals have the same origin as the differences between each first element and its heavier homologues in Groups 13 - 18 (see Section 3.2).

7.1 PROPERTIES OF HYDROGEN

7.1.1 The hydrogen molecule

Because of its high ionization energy hydrogen cannot form a metallic phase, except perhaps at extreme pressures ("atmospheres" of several planets have been suggested to consist of such material). Instead, the stable form of hydrogen under normal conditions is the covalently bonded diatomic molecule H_2, dihydrogen. Because of the very small radius of the hydrogen atom, the H_2 molecule shows a high enthalpy of dissociation, equal to 436 kJ mol^{-1}. This should be compared with the enthalpies of dissociation of the Li_2 and F_2 molecules, which are 108 and 159 kJ mol^{-1} respectively. Due to the high activation energy for dissociation, the H_2 molecule is chemically not particularly reactive at low temperatures. On the other hand at high temperatures the H_2 molecule eagerly reacts with most elements. The H_2 molecule occurs in two forms: ortho-hydrogen with parallel spins for the protons (↑↑) and para-hydrogen with antiparallel spins (↑↓). Para-hydrogen has the lower energy and is the only form at temperatures close to absolute zero. The concentration of ortho-hydrogen increases with increasing temperature until the ratio ortho/para of 3/1 is attained. Ortho- and para- hydrogen differ slightly in their properties, e.g. in their boiling points, and can be separated. Conversion between the two forms is a slow process because it is forbidden, being a transition between symmetric and antisymmetric rotational levels.

7.1.2 Reactions of hydrogen

Due to its intermediate electronegativity, hydrogen forms compounds with most elements, both electronegative and electropositive. The hydrogen atom can react in three ways:

1. The $1s$ valence electron may be lost to give the H^+ ion, i.e. the proton which, because of its very small size (about 1.5×10^{-13} cm) is always associated in condensed phases with other atoms or molecules. The most important adducts are the oxonium H_3O^+ and ammonium NH_4^+ ions.

2. The unpaired $1s$ electron of the hydrogen atom can form a covalent bond with an unpaired electron of another atom. Hydrogen forms covalent hydrides mainly with p-block elements, and the greatest contribution from covalence to bonding is observed for elements of similar electronegativity, i.e. for Group 14 and Group 15 elements. In the majority of cases the H–A bond has a partly polar character, wherein the hydrogen atom usually acquires a positive charge so that the H–A group becomes more or less acidic.

3. The hydrogen atom can acquire an electron to attain the $1s^2$ configuration of He and form the hydride ion H^-, with a radius which varies from 122 pm in LiH to 154

pm in CsH. Because the electron affinity of hydrogen is low, the acquisition of an electron is possible only in reactions with highly electropositive metals. These include the alkalis and the alkaline earths, except for Be and Mg with their rather high first and second ionization potentials. Low electron affinity is also responsible for the wide variability of the ionic radius of H^-. The reason is that low electron affinity makes it difficult for the hydrogen atom to retain the extra electron, with this difficulty increasing as the size of the cation decreases. The presence of the H^- ion in these salt-like hydrides is shown by the fact that on electrolysis of the molten salts hydrogen is evolved at the anode. The saline hydrides react vigorously with water with evolution of hydrogen. They, particularly CaH_2, are used in the laboratory to remove traces of water from solvents and from inert gases. This formation of salt-like hydrides in which hydrogen exhibits an oxidation number of -1 make it resemble the halogens.

Hydrogen shows some very specific properties:

1. It can act as a bridging atom between two boron atoms in boron hydrides (and indeed also between some d ions in binuclear complexes). The BHB bond is an electron deficient three-centre - two-electron (3c-2e) bond (Fig. 5.1). In the diborane molecule (B_2H_6), where boron shows sp^3 hybridization, there are two such bonds, shown schematically in Fig. 7.1.

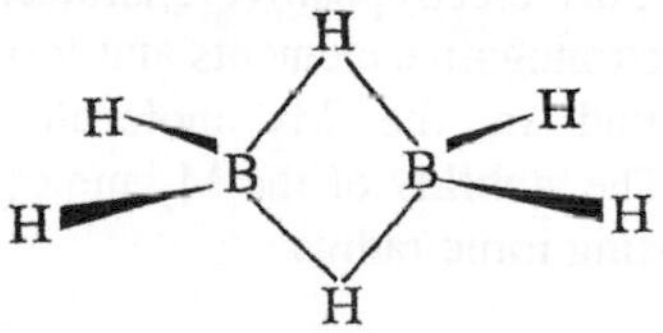

Fig. 7.1 Bridging of boron atoms by hydrogen atoms in the B_2H_6 molecule.

2. Because of its small size and intermediate electronegativity, hydrogen can act as a bridging atom between two highly electronegative atoms X and Y, each having a lone electron pair. The reaction which leads to formation of a hydrogen bond is that between the proton donor XH and the proton acceptor Y. The position of the hydrogen atom in the bond is usually nonsymmetrical and the bond, which can be described as a three-centre - four-electron bond (3c-4e, see Fig. 5.1), is rather weak. It is strongest when both X and Y are N, O or F atoms. Hydrogen-bonding plays a very important role in the chemistry of water and aqueous solutions, and in the chemistry of biopolymers.

3. Hydrogen forms hydrides with d-block metals of Groups 3 to 5, also palladium, and with f-block elements. These hydrides are as a rule non-stoichiometric, e.g. $ZrH_{1.75}$ or $LaH_{2.87}$, and show limiting stoichiometries which vary from MH to MH_3 for d-block elements, from MH_2 to MH_3 for lanthanides and actinides. The bonding in these hydrides is partly metallic and partly ionic, depending on the stoichiometry. The metallic conduction of these hydrides usually decreases with increasing hydrogen content. This is because electrons supplied by the hydrogen atoms are accommodated in the conduction band until this is filled. The Group 10 elements are used as hydrogenation catalysts, which implies formation of hydrides on the surface of the catalyst. However, at moderate pressures only palladium forms a stable PdH_x ($x < 1$) bulk phase. The explanation is that of the Group 10 elements

palladium has the lowest atomization enthalpy, which favours rupture of Pd–Pd bonding in the metal.

7.2 THE ALKALI METALS

7.2.1 General properties

The alkali metals have very low first ionization potentials, I_1. This results in high chemical reactivity, particularly with respect to electronegative elements. Reactivity increases down the Group i.e. with decreasing first ionization potential and electron affinity. For instance lithium reacts only slowly with water, whereas sodium reacts vigorously, potassium inflames, and rubidium and caesium react explosively. Reactivity towards liquid bromine also increases very markedly down the Group. Because of very high second ionization potentials, I_2, which range from 76 eV for Li to 23 eV for Cs, the alkali metals show only the oxidation number +1.

As they are electron-deficient the alkali metals can form only metallic solid phases. Their large metallic radii and small number of bonding electrons (two electrons per eight bonds in the case of a body-centered cubic lattice) result in small lattice energies and, therefore, low boiling and melting points, which decrease down the Group. If francium could occur in weighable amounts it would probably be a liquid under normal conditions. The highly electropositive character of the alkali metals means they react mainly with electronegative elements and form typical ionic compounds. Covalent bonding is found in the M_2 molecules and in the organometallic compounds of lithium. The stability of the M_2 molecules decreases from lithium to caesium, i.e. with increasing ionic radius.

7.2.2 Changes of properties down the Group

Fundamental properties of atoms such as $\langle r_{ns} \rangle$, r_i, r_{met}, ε_{ns} and I_1 (Table 7.1) change down the Group in a remarkable way, Fig. 7.2. In particular we notice:

– Relatively large changes in I_1, ε_{ns}, r_{met}, and r_i on going from lithium to sodium. These changes would have been greater if not for the fact that the $3s$ shell is built up from the very small $2p$ subshell of the neon atom, see Fig 7.3.

– Large changes in atomic and ionic parameters between Na and K and small changes in the K→Fr range. The reason is that the ns orbitals are built up from the $(n-1)p$ subshells in Ne, Ar, Kr, Xe and Rn. These shells differ considerably in their spatial extent between Ne and Ar but distinctly less in the Ar→Rn range, Fig. 7.3. Because of the relativistic effect in the $7s$ orbital, the position of francium with respect to orbital energy, atomic radius and ionization potential is between rubidium and caesium. It is seen in Fig. 7.2 that with respect to their fundamental atomic properties the alkali metals can be divided into three subgroups, viz:

Li　　　Na　　　K, Rb, Cs, Fr

This division is also valid for many chemical properties and is shown by e.g. solubilities in water of many salts, affinity for ion exchangers, enthalpy of hydration and enthalpy of dissociation of the M_2 molecules. Another example is afforded by the reaction with oxygen. When burnt in air Li forms Li_2O, K, Rb and Cs form

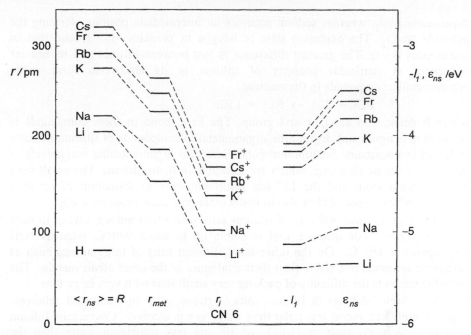

Fig. 7.2 Radii, energy of *ns* orbitals and first ionization potentials of Group 1 elements.

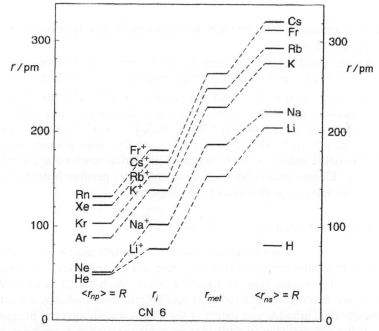

Fig. 7.3 Correlation between radii of Group 1 elements and radii of the (*n*−1)*p* subshells in noble gases (the 1*s* shell in He).

superoxides MO_2, whereas sodium occupies an intermediate position, forming the peroxide Na_2O_2. The oxidation state of oxygen in peroxides is -1, and that in superoxides is $-\frac{1}{2}$. The greatest difference is that between lithium and its heavier congeners. A particular property of lithium is its ability to form stable organometallic compounds in the reaction:

$$RCl + 2Li \rightarrow RLi + LiCl$$

where R denotes an alkyl or aryl group. The Li–C bond in these compounds is covalent to a significant degree; the organometallic compounds of lithium dissolve in liquid hydrocarbons. Sodium and potassium form organometallic compounds of much greater ionic character, which are insoluble in hydrocarbons. The small radii of the lithium atom and the Li^+ ion result not only in formation of covalent organometallic compounds but also in many other particular properties, e.g.:

– The high thermal stability of lithium salts with small anions, owing to high lattice energies. For instance LiH is stable up to about 900 °C, whereas NaH decomposes at 350 °C. On the other hand, lithium salts of large anions such as carbonate are much less stable than their analogues of the other alkali metals. The probable reason is the difficulty of packing very small ions with very large ions.

– Very high solubility of lithium salts of strong acids in water and relatively high solubility in solvents less polar than water as e.g. alcohols. Conversely, lithium salts of weak acids such as H_2CO_3 or HF are less soluble in water than the respective salts of heavier congeners.

– Formation of lithium-bonds, the analogues of hydrogen-bonds, in $HF \cdots LiF$ and $LiF \cdots LiF$;

– Binding of dinitrogen from air with the formation of ruby-red crystalline Li_3N (diagonal similarity to Mg).

Although lithium is not the first element in the Group it shows these and other unique properties because its $2s$ shell is built over the very small $1s$ shell.

Whereas lithium is so different from the rest of Group 1 elements, potassium, rubidium and caesium show many similarities e.g. with respect to low solubility in water of salts with large anions such as ClO_4^-, $[PtCl_6]^{2-}$, or $[Co(NO_2)_6]^{3-}$. Another example is the formation by the large K^+, Rb^+, and Cs^+ ions of ionic salts with several rather unstable anions, such as the (large) polyhalide anions and a number of fluorometallates. Sodium usually occupies an intermediate position between lithium and its heavier homologues with respect to its properties.

7.2.3 Structures of salts and aqueous solution chemistry

The predominantly ionic character of most salts of the alkali metals means that the majority have simple structures determined by size and electrostatics. Thus most of the halides have the sodium chloride structure, where each cation is octahedrally surrounded by six anions, and each halide is octahedrally surrounded by six cations. However the chloride, bromide and iodide of the large caesium ion (and presumably francium – if it were available in sufficient amount) have an 8:8 structure rather than the 6:6 sodium chloride structure. This difference in structure is in accordance with expectations from radius ratios, as the caesium halides mentioned have radius ratios larger than the value of 0.732 which marks the borderline between the 6 : 6 NaCl and 8 : 8 CsCl structures.

The large ionic radius and small charge of the alkali metal cations result in relatively small negative enthalpies of hydration, whose magnitude decreases going down Group 1 (Table 7.1). This decrease shows the characteristic pattern

Li   Na   K, Rb, Cs

documented in the previous section. ΔH_{hydr} correlates linearly with $1/r_i$; it is the irregular change in r_i which causes the division into the three sub-groups. The number of water molecules in the first coordination sphere of the cation increases from 4 for Li^+ to 8 for Cs^+, but the total number of water molecules hydrating the cation decreases on going from Li^+ down to Cs^+. In spite of low hydration enthalpies which stabilize the cations to a rather small extent, the standard reduction potentials, $E^o(+1/0)$, i.e. potentials for the process $M^+(aq) + e^- \rightarrow M(s)$, are strongly negative (see Table 7.1) due to the very low first ionization energies. Values of between -2.71 and -3.04 V for $E^o(+1/0)$ may be compared with $E^o(+2/0)$ between -2.36 and -2.91 V for the alkaline earth (Group 2) metals Mg to Ba, -0.76 V for Zn, and $+0.34$ V for Cu.

7.2.4 Solutions in liquid ammonia

In spite of the low solvation energies of M^+ cations, small atomization energies (weak metallic bonding) and very low first ionization energies permit the alkali metals to dissolve in liquid ammonia and to form solvated 1+ cations:

$$Na(s) + NH_3(l) \rightarrow Na^+(am) + e^-(am)$$

The dissolution process is promoted by solvation of both cation and electron by the NH_3 molecules. Ammonia solutions of the alkali metals are paramagnetic and colored, and are strong reductants. They are, for instance, able to reduce metallic gold to $Au^-(am)$. Similar, but less stable, are solutions of alkali metals in amines, tetrahydrofuran and ethers. Sodium also forms an ionic adduct with aromatic hydrocarbons e.g $Na^+(anthracene)^-$. Because of their small electron affinities, the alkali metals generally do not tend to fill the s subshell by acquiring an electron and forming the M^- anion. However, solutions of alkali metals, except Li, in ammonia and ethers may contain not only solvated M^+ cations and solvated electrons but also M^- anions. On addition of cryptands or crown ethers to complex and thus stabilize the cation, crystalline solids are formed. Some of these contain M^- anions and are called alkalides, others contain trapped electrons and are called electrides. An example of the former is the salt $[Na([2,2,2])]^+Na^-$ where sodium exhibits oxidation numbers of both $+1$ and -1. An example of an electride is the dark blue paramagnetic solid of formula $[Cs(18\text{-crown-}6)]^+e^-$ in which the counter-anion is the electron.

7.2.5 Coordination chemistry and separation

Cations of the alkali metals are extremely weakly complexed by most potential ligands, uncharged and anionic. In order to obtain complexes of significant stability it is necessary to use ligands specifically designed to maximize stabilities. The alkali metal cations are, in Pearson's Hard and Soft Acids and Bases classification, "hard". It is therefore advisable to select potential ligands with "hard" donor atoms, such as oxygen or nitrogen. However, simple monodentate ligands do not form complexes of useful stability with alkali metal cations. Other effects, such as the chelate effect, the macrocyclic effect, and encapsulation, have to be employed to increase stabilities significantly.

Stability is considerably enhanced if a ligand has more than one donor atom bonded to a given metal ion. Thus complexes of 1,2-ethanediamine are markedly more stable than their bis-ammine, $(NH_3)_2$, and bis-amine, $(RNH_2)_2$, analogues, with further increase in stability on going from $H_2NCH_2CH_2NH_2$ to $H_2NCH_2CH_2NHCH_2CHNH_2$ and so on. Further, the stability of a complex of a macrocyclic ligand is higher than that of its linear analogue – the ligand

$$
\begin{array}{cc}
HNCH_2CH_2NH \\
| \qquad\qquad | \\
H_2C \qquad\quad CH_2 \\
| \qquad\qquad | \\
H_2C \qquad\quad CH_2 \\
| \qquad\qquad | \\
HNCH_2CH_2NH
\end{array}
$$

forms more stable complexes than $H_2NCH_2CH_2NHCH_2CH_2NHCH_2CH_2NH_2$. The former phenomenon is termed the chelate effect, the latter the macrocyclic effect. In both cases it is actually very difficult to make exact comparisons, for in both cases the ligand comparisons are not direct. The chelating ligand $H_2NCH_2CH_2NH_2$ has two primary amine donors, whose σ-donor strength is significantly different from that of two ammonias. Similarly the macrocyclic ligand shown above has four secondary amine donors, whereas its linear analogue has two primary and two secondary amine donor groups. The apparent magnitude of these effects also depends on the standard states (concentration units) used in defining their stability constants. Nonetheless both effects are real and substantial under normal laboratory conditions. In the present context it can be stated that, apart from the special case of the solvated ions in solution in water, liquid ammonia, alcohols, and other donor solvents, the alkali metal cations do not form stable complexes with monodentate ligands. They form complexes, albeit of low stability, with bidentate O-donor ligands such as β-diketones and salicaldehyde, but to obtain complexes of significant stability in solution one must use macrocyclic ligands such as the crown ethers shown in Fig. 7.4. Some of these complexes are stable in non-aqueous media, but to obtain complexes stable in aqueous solution it is necessary to go beyond the macrocyclic effect and to encapsulate the alkali metal cation in a cryptand such as those shown in Fig. 7.5.

These crown ether and cryptand complexes of alkali metal cations are of particular interest in view of the great importance of size effects, specifically of matching the radius of the cation to the radius of the ligand cavity. This is demonstrated in Fig. 7.6 for a series of cryptands of varying cavity size. This size matching is also of great importance in biological systems, where it forms the basis of the selective transport of sodium and potassium ions across membranes.

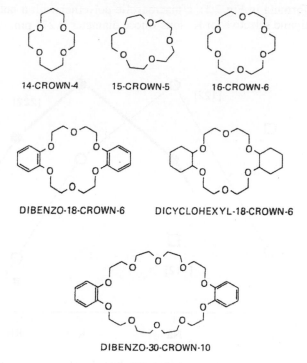

Fig. 7.4 Crown ethers, with their conventional names.

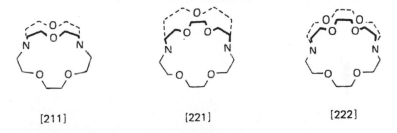

Fig. 7.5 Cryptands, with their conventional designations – the numbers within the square brackets indicate the value of n for each of the $-CH_2(CH_2OCH_2)_nCH_2-$ links.

Since most of the chemistry of the alkali metals is that of M^+ cations and complexes are weak, separation is mainly based on differences in the size of either hydrated or non-hydrated cations. The respective examples are column chromatography using either a strongly acidic cation exchanger or column chromatography with a tailor-made inorganic ion exchanger or extractant. The latter processes show much higher selectivity. In particular, effective separation can be achieved by solvent extraction using ligands with the binding site in the form of a hole fitting the size of one of the ions to be separated. Thus, for example, dibenzo-

18-crown-6 (formula in Fig. 7.4), a macrocyclic polyether with a hole size of about 290 pm, is a ligand selective for K^+ ions whose diameter is 276 pm.

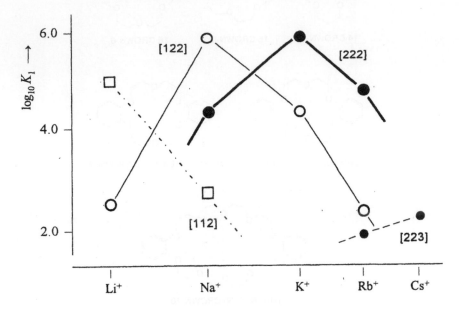

Fig. 7.6 Stability constant trends for cryptates of alkali metal cations. {Note that the x axis scale is arbitrary, as the coordination number of the cation is not known in many cases and hence, in view of the variation of ionic radius with coordination number, it would not be possible to position data points with confidence on the x axis scale.}

8

Group 2. The alkaline-earth metals

Table 8.1 — Fundamental properties

	Be	Mg	Ca	Sr	Ba	Ra
$R = \langle r_{ns} \rangle$ / pm	140	172	222	241	269	266
I_1 / eV	9.32	7.64	6.11	5.69	5.21	5.28
I_2 / eV	18.2	15.0	11.9	11.0	10.0	10.1
χ	1.57	1.31	1.00	0.95	0.89	0.89
mp (°C)	1280	650	839	769	729	700
r_{met} / pm	113	160	197	215	217	220
r_i / pm CN 4	27	57				
CN 6	45	72	100	118	135	~ 140
CN 8		89	112	126	142	148
CN 12			134	144	161	170
ΔH_{hydr} / kJ mol^{-1} a	−2492	−1931	−1584	−1452	−1314	−1303
$E°(+2/0)$ / V	−1.97	−2.36	−2.86	−2.90	−2.91	−2.92

a Based on ΔH_{hydr} (H^+) = − 1094 kJ mol^{-1}

8.1 GENERAL PROPERTIES

The Group 2 elements, also called the alkaline-earth metals, have rather low I_1 and moderate I_2 values. This results in high chemical reactivity, particularly with respect to electronegative elements, including reaction of Be and Mg with the strongly bonded N_2 molecule. Because of relatively low ionization potentials and high solvation enthalpies of the M^{2+} ions, the elements from magnesium to barium, like the alkali metals, dissolve in liquid ammonia. Since the Group 2 elements are, except for beryllium, highly electropositive they form ionic compounds with most elements, in which they exhibit the oxidation number +2. A large covalency contribution is observed in most beryllium compounds and in organomagnesium compounds, such as MgR_2 and RMgX (Grignard reagents). Because of their large radii, organometallic compounds of Ca, Sr and Ba are highly ionic and reactive. As is the case with alkali metal hydrides the Group 2 hydrides MH_2 contain the H^- anion, except for BeH_2 and MgH_2 where there is a high contribution from covalency to the bonding. The Group 2 elements do not show the +1 oxidation state. This is mainly due to the large ionic radii of the hypothetical M^+ ions, which would result in relatively low hydration enthalpies (see Section 5.2) and lattice energies. Because of their closed ns subshell, the alkaline-earth metals do not form M_2 molecules and have very low electron affinities. These at first decrease from -0.19 eV for Be to -1.9 for Ca and then increase to -0.48 for Ba. In respect to their low electron affinities, the Group 2 elements resemble the elements which have closed p and d subshells, see Fig. 4.3.

As they are electron-deficient the alkaline-earth elements can form only metallic solid phases. Since the ns subshell is filled, formation of a metallic phase is possible only because the np orbitals participate in the energy band. The alkaline-earth elements have highly negative standard reduction potentials $E^{\circ}(+2/0)$. This is because the sum of ionization energies $I_1 + I_2$ is effectively opposed by the relatively large negative enthalpies of hydration of the M^{2+} ions.

8.2 CHANGES OF PROPERTIES DOWN THE GROUP

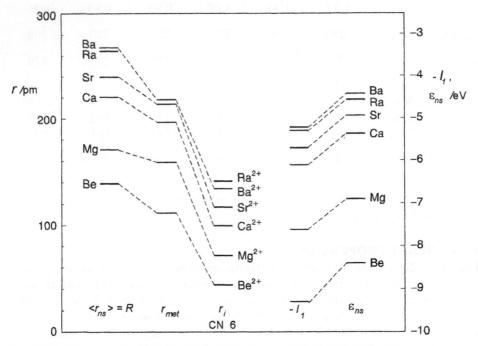

Fig. 8.1 Radii, energy of ns orbitals and first ionization potentials of Group 2 elements.

Changes of chemical properties going down the Group result from changes in R (= $\langle r_{ns} \rangle$), r_i, r_{met}, ε_{ns} and $I_1 + I_2$. These are, as a rule, irregular (Fig. 8.1). In particular we observe:

– Large changes in I_1, ε_{ns}, r_{met} and r_i between beryllium and magnesium. The reason is the same as that found in the case of lithium and sodium.

– Large changes in atomic and ionic parameters between Mg and Ca and small changes in the sequence Ca → Ra. This behaviour again parallels changes for the Group 1 elements and has the same origin, i.e. building the ns orbitals over the $(n-1)p$ subshells in the noble gases Ne to Rn. The $(n-1)p$ subshells differ considerably in their spatial extent between Ne and Ar and distinctly less in the sequence Ar → Rn, Fig. 8.2.

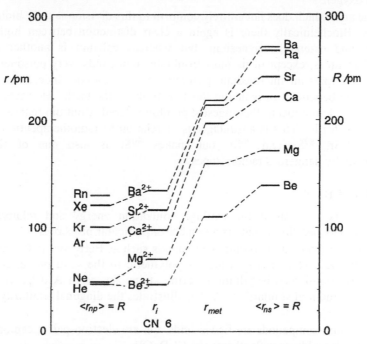

Fig. 8.2 Correlation between radii of Group 2 elements and radii of the $(n-1)p$ subshells in noble gases (the $1s$ shell in He).

– Because of the relativistic effect in the $7s$ orbital, the position of radium with respect to orbital energy, ionization potential and atomic radius is between strontium and barium.

– There is an unusually large variation in coordination numbers, in simple compounds and in complexes, ranging from almost universal four-coordination for beryllium to 12-coordination in certain compounds of the heavier elements (cf. ionic radii shown in Table 8.1).

It is seen in Fig. 8.1 that with respect to fundamental atomic properties the alkaline-earth metals can be divided into three subgroups:

Be   Mg   Ca, Sr, Ba, Ra

This division finds its counterpart in many chemical properties, e.g. in enthalpies of hydration, standard reduction potentials, affinities of the cations for ion exchangers, structures and coordination numbers, and solubilities of salts in water. For example, certain salts of strontium, barium, radium and, to some extent, calcium, including sulphates and phosphates, are sparingly water-soluble or insoluble, those of magnesium and beryllium significantly more soluble. The division into three subgroups is also shown by the formation of peroxides. Beryllium does not form a peroxide, magnesium only a hydrated peroxide, but calcium, strontium, and barium form relatively stable peroxides. However these three peroxides all decompose at higher temperatures, yielding oxygen. Indeed BaO_2 was once used as a renewable source of oxygen.

The one area which does not entirely conform to this division is their bioinorganic chemistry. Biochemically there is again a clear distinction between highly toxic beryllium and essential magnesium, but whereas calcium is another essential element barium is, except in its most insoluble compounds, very poisonous, while strontium occupies an intermediate position. It is non-toxic, indeed its salts are claimed to be beneficial in reducing the sensitivity of the teeth. However its close similarity to calcium means that ingested or administered strontium readily localizes in bones and teeth. This is advantageous in relation to radiotherapeutic use of its isotopes ^{87m}Sr, ^{89}Sr and ^{90}Sr, but makes ^{90}Sr is also one of the most environmentally dangerous radioisotopes.

8.3 BERYLLIUM

Because of its very small radius, high ionization energy and relatively high electronegativity beryllium exhibits a number of particular properties:

– Chemical bonds in beryllium compounds such as BeH_2 and $BeCl_2$ are highly covalent. In the case of $BeCl_2$, which is polymeric in the solid phase, containing tetrahedrally coordinated beryllium, chlorine atoms act, as in Al_2Cl_6, as bridging atoms. This structural similarity to Al_2Cl_6 illustrates the diagonal similarity of Be to Al.

– Beryllium compounds are Lewis acids, i.e. are electron-pair acceptors. $BeCl_2$ forms molecular adducts with ethers, e.g. Cl_2BeOR_2.

– In contrast to the other alkaline earth metals, beryllium does not react with water.

– Metallic beryllium dissolves in aqueous NaOH solutions with evolution of hydrogen, and $Be(OH)_2$ is amphoteric – further examples of diagonal similarity to aluminum.

– Beryllium compounds are highly poisonous, probably because they form strong covalent bonds to nitrogen atoms in proteins, which inhibits their enzymatic functions.

– The standard reduction potential $E^°(+2/0)$ for this first element in the Group is significantly less negative than that for the heavier elements. That is the opposite to what is observed for the alkali metals. The reason is that the high values of I_1+I_2 and of atomization (vaporisation) energy for beryllium predominate over the enthalpy of hydration of Be^{2+}.

These and other unique properties of beryllium show it to behave as the first member of the Group. This behaviour and position are, however, inconsistent with the assignment of helium as the first member of Group 2 on the basis of shell structure. The uniqueness of beryllium, and also of lithium, cannot result from the lack of an inner shell with the same quantum number l (see Section 3.2), because the atoms of both elements have the $1s$ inner shell. Rather these two elements owe their singular properties to building the $2s$ shell directly over the very small and tightly bound $1s$ shell.

8.4 STRUCTURAL CHEMISTRY

Apart from beryllium, the structural chemistry of compounds of the Group 2 elements is very much in accordance with their predominantly ionic character. Thus the oxides MgO → BaO and the sulphides MgS → BaS all have the sodium chloride structure, with both cation and anion in octahedral coordination – BeO and BeS contain tetrahedrally coordinate Be, in the wurtzite and zinc blende structures respectively. Just as in the case of the alkali metal halides (Section 7.2.3), the structures of the alkaline earth dihalides depend on the relative sizes of cation and anion – the respective typical structures are summarized in Table 8.2.

Table 8.2 – Ionic environments in simple structures.

		CN ratio cation : anion	cation environment	anion environment
MX	sodium chloride	6 : 6	octahedral	octahedral
	caesium chloride	8 : 8	cubic	cubic
MX_2	fluorite (CaF_2)	8 : 4	cubic	tetrahedral
	rutile (TiO_2)	6 : 3	octahedral	triangular

CaF_2, SrF_2 and BaF_2 have the fluorite structure, with the cation surrounded by eight anions, MgF_2 has the rutile structure, while BeF_2 contains, as ever, four-coordinated Be. $SrCl_2$ and $BaCl_2$, but not $CaCl_2$, can adopt the fluorite structure; $CaCl_2$ and the bromides and iodides of Sr and Ba adopt more complicated structures, still with the high coordination numbers of 7 or 8, to accommodate the larger metal cations better.

The structural chemistry of the chlorides is complicated by their affinity for solvents, particularly water. Calcium chloride is a long-established, if not particularly efficient, desiccant, forming hydrates $CaCl_2.2H_2O$, $CaCl_2.4H_2O$ and $CaCl_2.6H_2O$, and also an ammoniate $CaCl_2.8NH_3$. The hexahydrates of the chlorides, bromides and iodides of calcium and strontium all adopt the so-called $SrCl_2.6H_2O$ structure, which contains infinite chains of cations linked by bridging water molecules – overall each cation is surrounded by nine water molecules. In contrast, several hydrates of magnesium and beryllium salts contain the simple entities $[Mg(H_2O)_6]^{2+}$ or $[Be(H_2O)_4]^{2+}$ which behave as spherical cations.

Examples include $MgCl_2.6H_2O$, $Mg(ClO_4)_2.6H_2O$ and $BeSO_4.4H_2O$. $[Mg(H_2O)_6][ReCl_6]$ has the sodium chloride structure, since cation and anion are of suitable sizes to pack in this 6:6 structure.

Salts of non-spherical polyatomic anions may also have relatively simple structures, often based on the sodium chloride structure. Thus, for example, calcium carbide, CaC_2, has the NaCl structure, elongated along the z axis to accommodate the C_2^{2-} anions, which give ethyne, $HC{\equiv}CH$, on hydrolysis (contrast methane from Al_4C_3 – see Section 10.1).

8.5 COORDINATION AND SOLUTION CHEMISTRY

Because of their large ionic radii and low hardness (except Be and, to a certain degree, Mg), alkaline earth metal cations do not in general form complexes with simple ligands. Only Be^{2+} forms stable anionic complexes with halogens, e.g. BeF_4^{2-}. Magnesium forms fairly stable complexes with several common bidentate ligands; the magnesium and calcium cations form complexes with polydentate ligands such as EDTA { $(O_2CCH_2)_2NCH_2CH_2N(CH_2CO_2)_2^{4-}$ }. Ca^{2+}, Sr^{2+} and Ba^{2+} form complexes with crown ethers and cryptands. Nitrogen ligands generally form weak complexes, with the notable exception of porphine derivatives which form stable complexes with magnesium, for instance chlorophyll. The characteristic coordination number for beryllium is four, as in BeF_4^{2-} and, e.g., the aqua cation $[Be(H_2O)_4]^{2+}$. The coordination number for Mg^{2+} is almost invariably 6, for Ca generally 6 or 8, while Sr^{2+}, Ba^{2+} and Ra^{2+} ions usually have coordination numbers of 7 or 8. Because of their similar r_i values, the Sr^{2+}, Ba^{2+} and Ra^{2+} ions show similar affinities for strongly acidic cation exchange resins and are, therefore, hard to separate in this way. However, separation can easily be accomplished on certain inorganic ion exchangers such as crystalline polyantimonic acid, selective for Sr^{2+} ions, or on crystalline manganic acid which makes separation of radium from barium possible.

In aqueous solution the Mg^{2+} ion has a hydration number of six, and a solvation number of six in several non-aqueous solvents. This has been demonstrated by n.m.r. spectroscopy, whose time-scale has permitted the determination of kinetic parameters for solvent exchange. The residence time for water molecules in the primary hydration shell of Mg^{2+}aq is about 10^{-6} s at ambient temperatures. Water exchange is about a thousand times slower at Be^{2+}aq, whose much higher enthalpy of hydration (-2492 kJ mol^{-1}; cf. -1931 kJ mol^{-1} for Mg^{2+}aq) indicates stronger cation–water bonding, consistent with the considerably smaller radius of Be^{2+}. Hydration numbers are not known with certainty for Ca^{2+}, Sr^{2+}, Ba^{2+} and Ra^{2+}, but are probably eight; water exchange with their primary hydration shells is fast ($k >$ 10^8 s^{-1}).

The kinetics of complex formation from Mg^{2+}aq in aqueous solution have been much studied. Dissociative activation is indicated for complex formation, as for solvent exchange. This is as expected for a small ion with a fairly strongly bonded primary solvation shell. Rather little is known about kinetics and mechanisms of solvent exchange and complex formation at Ca^{2+}, Sr^{2+} and Ba^{2+} – these reactions are rapid and are technically difficult to monitor.

The formation of ternary complexes of Ca^{2+}, Sr^{2+} and Ba^{2+}, and of Mg^{2+}, is a key step in a number of template reactions. These are important synthetically for the production of a range of macrocyclic compounds, particularly crown ethers. The relative effectiveness of the various cations depends on size factors, the cation whose radius most closely matches the hole size of the developing macrocycle being the best template. Thus Ba^{2+}, whose radius is close to that of K^+, is an effective template for the assembly of 18-crown-6, Fig. 8.3. The Group 2 cations are also effective templates for the formation of mixed nitrogen/oxygen-containing macrocycles. Reaction of 3,6-dioxo-1,8-diamino-octane, dodao, with 2,6-diacetylpyridine, dap, on Ba^{2+} (or the similarly large Pb^{2+}) as template involves condensation of two molecules of each to give the 30-membered ring macrocycle shown as (a) in Fig. 8.4. If Sr^{2+} is used as template, then reaction only proceeds as far as the dap + 2dodao intermediate, Fig. 8.4(b). Ca^{2+} is ineffective, but the use of Mg^{2+} leads to the 15-membered ring 1dap + 1dodao product, Fig. 8.4(c), albeit in low yield. Mg^{2+} is also an effective template for the production of the 14-membered ring of Fig. 8.5(a) from dap and hydrazine, of a 15-membered ring (Fig. 8.5(b)) from dap and triethylenetetramine. Here, and in the template generation of porphyrin derivatives, Mg^{2+} favours the formation of small ring polyaza-macrocycles – the larger alkaline earth cations favour 18-membered, and larger, rings and macrocycles containing both oxygen and nitrogen, or solely oxygen, hetero-atoms.

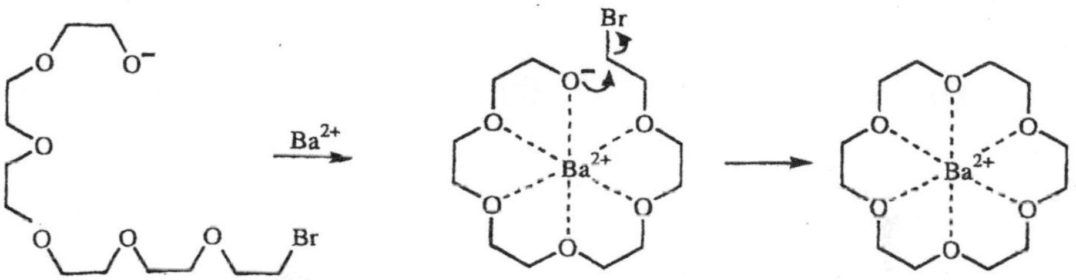

Fig. 8.3 Ba^{2+} as template for the formation of 18-crown-6.

Fig. 8.4 The effect of metal ion size on the products of template reactions.

dap

dodao

Ba^{2+} Sr^{2+} Mg^{2+}

(a)   (b)   (c)

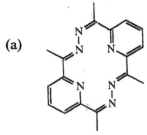

(a)

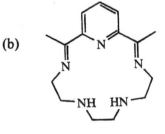

(b)

Fig. 8.5 Macrocyclic rings obtained on Mg^{2+} templates.

9

Groups 13 and 3

We shall devote the main part of this Chapter to the Group 13 elements boron to thallium, but in the final section deal briefly with the Group 3 elements scandium, yttrium and lanthanum, which form a link between aluminium and the lanthanides (Chapter 18).

9.1 THE GROUP 13 ELEMENTS

Table 9.1 — Fundamental properties of the Group 13 elements

	B	Al	Ga	In	Tl
$R = \langle r_{np} \rangle$ / pm	117	181	179	194	186
I_1 / eV	8.30	5.98	6.00	5.79	6.11
$I_2 + I_3$ / eV	63.0	47.3	51.2	46.9	50.2
χ	2.04	1.61	1.81	1.78	2.04
mp (°C)	2300	660	29.8	156	303
r_{met} / pm	88 [a]	143	135	167	170
$r_i(1+)$ / pm (CN 6)				~ 120	150
$r_i(3+)$ / pm (CN 4)	11	39	47	62	75
(CN 6)		53.5	62.0	80.0	88.5
$E°(3+/0)$ / V		−1.68	−0.53	−0.34	+0.72
$E°(3+/1+)$ / V				−0.44	+1.25

[a] r_{cov}

Table 9.1 and Fig. 9.1 show that with respect to many fundamental atomic properties, boron differs markedly from the rest of the Group 13 elements. Boron is also chemically different, resembling silicon rather than its own heavier congeners – an example of diagonal similarity. The remarkable chemical properties of boron are the result of high ionization energies, small atomic radius, relatively high electronegativity and unavailability of d orbitals. High ionization energies and small radius result, in turn, from the unique properties of the $2p$ shell (see Figs. 3.1 and 4.1).

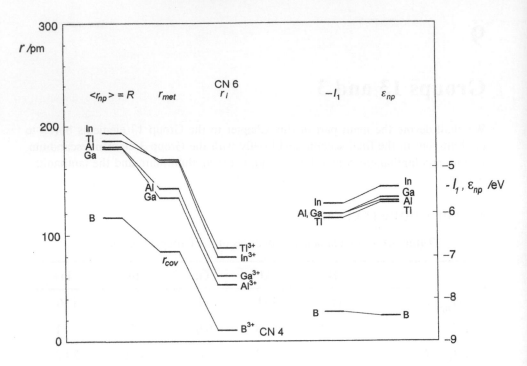

Fig. 9.1   Radii, energies of *np* orbitals and first ionization potentials of Group 13 elements.

9.2 PROPERTIES OF BORON

With the number of valence electrons less than twice the number of valence orbitals boron could form a metallic phase. However, because of high ionization potentials boron prefers to form covalent solid phases. On the other hand because boron has more valence orbitals than electrons, i.e. is electron deficient, it cannot form a three-dimensional array of σ-bonded atoms like that shown by carbon in diamond, and must adopt an intermediate structure based on cluster formation. Therefore, the structural unit in all allotropic forms of elementary boron is the B_{12} icosahedron (cluster) where bonds between boron atoms are covalent. Bonding within the icosahedron is best described in the framework of the molecular orbital theory. According to this approach the 26 valence electrons in the cluster fill the 13 bonding molecular orbitals. In the simplest allotrope the 10 remaining electrons serve to link the icosahedra by 2c-2e and 3c-2e bonds in a manner similar to cubic closest packing of spheres. Elementary boron is a poor conductor of electricity (though it does conduct at high temperatures), has a large atomization enthalpy (504 kJ mol^{-1}) and very high melting temperature. The reason for the latter properties is the very small radial extent of the 2s and 2p orbitals, which makes the distance between the boron atoms in the B_{12} cluster short (about 176 pm), hence bonding strong. Because of its high ionization potentials I_1, I_2 and I_3 boron is also extremely inert chemically. It is unaffected by HCl or HF, and attacked only with difficulty by hot HNO_3.

Boron forms binary borides M_nB_m with many metals which show great diversity of stoichiometries and structural types. Metal-rich borides are very hard, chemically inert materials with high melting points and high electrical conductivities. Boron nitride, BN, is a particularly interesting compound in that it forms allotropes with the diamond and graphite structure (the BN group is isoelectronic with the CC group). Another interesting compound is borazine, $B_3N_3H_6$, which is isoelectronic and isostructural with benzene. Boron forms $p\pi-p\pi$ bonds with carbon and nitrogen as e. g. in the graphite form of BN and in borazine.

Boron forms covalently bonded molecules with the halogens; BF_3 is a gas and BCl_3 a liquid. The covalent character of B–X bonds and small B–X distances arise from the relatively high electronegativity of boron and donation of nonbonding electrons on the halogen atoms into the vacant p_z orbital of boron, which in the trihalides shows sp^2 hybridization. The donation of electrons from the ligands hinders formation of dimers by boron trihalides, contrary to trihalides of the heavier Group 13 elements which complete their valence octet this way. In BX_3 compounds the boron octet is incomplete. To make up the deficiency of electrons, boron trihalides are strong Lewis acids (electron acceptors) and give adducts with electron pair donors e.g. with ethers, with many nitrogen bases, or with halide ions; Et_2OBF_3, Me_3NBF_3 and BF_4^- can serve as examples. Formation of adducts is accompanied by a change from sp^2 to sp^3 hybridization. Either the unavailability of d orbitals or its small ionic radius (or both) preclude the formation of B^{III} complexes with coordination numbers higher than 4.

The simple hydrated B^{3+} cation is not known. B_2O_3, like SiO_2, shows acidic properties and on reaction with water gives the very weak acid H_3BO_3 (diagonal similarity to silicon). Boric acid is not so much a Brønsted acid (proton donor) as a Lewis acid, because it accepts the OH^- ion and gives the complex anion $[B(OH)_4]^-$. In salts of boric acid, borates, the structural unit is either $[BO_3]^{3-}$ or $[BO_4]^{5-}$. As in the case of SiO_4 tetrahedra, the triangular BO_3 and tetrahedral BO_4 units can share oxygen atoms and give polyborates, which show glassy properties.

A unique property of boron is the formation of boranes (boron hydrides) e.g. B_2H_6, B_4H_{10}, B_5H_9, B_5H_{11}, borane anions, e.g. $B_{10}H_{10}^{2-}$, and carboranes, e.g. $C_2B_{10}H_{12}$, most of which are cluster compounds (Fig. 9.2). Apart from the usual 2c–2e bonds many of these compounds also incorporate 3c–2e bonds to make up for the deficiency of electrons in the boron atom. For instance in the simplest borane, B_2H_6, there are four 2c–2e bonds (the terminal B–H bonds) and two 3c–2e bonds (the B–H–B bonds) – see Fig. 7.1. Neutral boranes, borane anions and carboranes show very complex structures that can be classified into three main series: *closo-*, *nido-* and *arachno-*boranes (and carboranes). Carboranes can be formally derived from boranes by replacing the BH group by the isoelectronic C atom. The *closo* (cave) boranes are closed deltahedral clusters (polyhedra with only triangular faces) which consist of n boron skeletal atoms and have the general formula $B_nH_n^{2-}$. The simplest example is the *closo*-borane anion, $B_6H_6^{2-}$, which consists of an inner octahedron of boron atoms surrounded by an outer octahedron of radially disposed

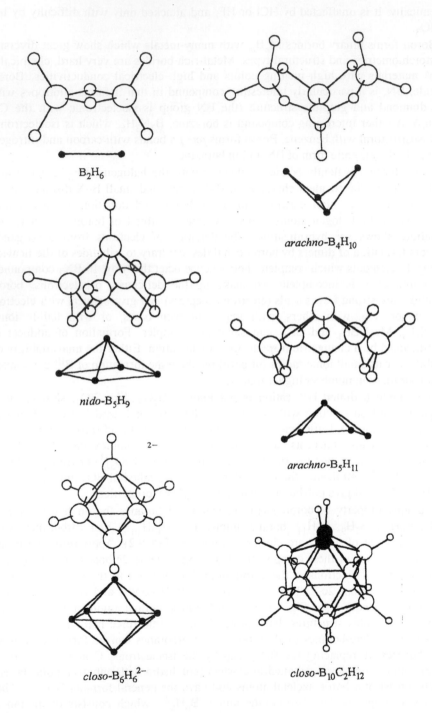

B_2H_6

arachno-B_4H_{10}

nido-B_5H_9

arachno-B_5H_{11}

closo-$B_6H_6^{2-}$

closo-$B_{10}C_2H_{12}$

Fig. 9.2 Formulae and structures of some boranes and a carborane.

hydrogen atoms. Formal removal of one vertex (the BH group with its two skeletal electrons) from this *closo* parent cluster and addition of 2 electrons and 4 protons yields an open *nido* (nest) cluster B_5H_9 (the general formula of neutral *nido* clusters is B_nH_{n+4}). Conceptually it is also possible to remove two vertices from $B_6H_6^{2-}$ and replace them by 4 electrons and 6 protons, which yields the even more open *arachno* (spider's web) cluster B_4H_{10} (the general formula is B_nH_{n+6}). The additional protons are necessary to satisfy electroneutrality and bonding requirements of the boron atoms which in the preceding compound were bonded to the detached BH group. In closed clusters, in contrast to open clusters, there are no bridging hydrogen atoms i.e. no 3c-2e B–H–B bonds. The open boranes, in addition to the B–H–B bonds, also frequently incorporate 3c-2e B–B–B bonds. For instance, in the *arachno*-borane, B_5H_{11}, there are two such B–B–B bonds. However, it should be noted that for larger boranes it becomes increasingly difficult to describe their structure in terms of two-centre and three-centre bonds – a molecular orbital treatment encompassing the whole molecule is required.

Boranes show low or moderate thermal stability mainly because of strong bonds in the parent elemental boron and in the H_2 molecule. Boranes are very reactive and the lower members are spontaneously inflammable in air. They are hydrophobic and readily dissolve in organic solvents. In contrast, the borane anions are stable and their alkali metal salts dissolve in water. The extensive chemistry of boron hydrides finds no parallel in the heavier elements of Group 13. The probable reason is the small size of the B atom and its electronegativity, almost equal to that of the H atom.

To recapitulate, most properties of boron place it rather on the non-metal side of the borderline between metals and non-metals. Along with carbon and hydrogen, boron shows the greatest difference between the first and next elements in any main Group of the Periodic Table.

9.3 PROPERTIES OF Al, Ga, In, and Tl

9.3.1 General

Aluminium and its congeners gallium, indium and thallium have much lower ionization potentials than boron. Hence they are genuine metals and are much more chemically reactive. Their melting points are low and change irregularly down the Group. Gallium has a very low melting point (29.8 °C), probably because it tends to form Ga_2 molecules in the metallic phase. Formation of Ga_2 molecules, or more strictly the presence of one short (and six longer) Ga–Ga distances in the metallic phase, may well be caused by the presence of the filled $3d$ shell. This stabilizes the $4s$ electrons, thereby making the second s electron somewhat reluctant to enter the energy band. The second s electron on each Ga atom may then participate in formation of the Ga–Ga bond. With its high boiling temperature of 2300 °C gallium shows the widest known temperature range for a liquid. Stabilization of the $4p$ and $4s$ electrons by the filled $3d$ shell also makes the ionization potentials I_1, I_2 and I_3 of gallium equal to, or higher than, the respective values for aluminium. In accordance with the inert pair effect Al, Ga, In and Tl show oxidation numbers +3 and +1. The stability of the higher oxidation state decreases down the Group (see

Section 5.2), of the lower increases enormously from negligible importance for Al to being the stable oxidation state of thallium.

There are several halides whose stoichiometric formula MX_2 (M = Ga, In, Tl; X = Cl, Br) suggests the +2 oxidation state. However these compounds contain gallium, indium or thallium in a stable +1 oxidation state, being mixed valence complex salts $M^I[M^{III}X_4]$. Gallium and indium do show the "forbidden" oxidation state +2, but only in dimeric ions with metal–metal bonds, as for example in $[Br_3Ga-GaBr_3]^{2-}$.

There is an important series of semiconducting materials which are binary compounds of the Group 13 elements Al, Ga, and In with the Group 15 elements P, As, and Sb. These 1:1 compounds have the zinc blende structure, i.e. both elements are in tetrahedral coordination, with the band gap decreasing the lower the constituent elements are in the Periodic Table. There is a 20-fold decrease in band gap on going from AlP down to InSb. One should note that in the 1:1 compounds of the Group 13 and Group 15 elements the number of electrons in the bonding band is 4 per atom, i.e. the same as in Si and Ge.

The M^{3+} ions of Group 13 elements, except for boron, can form complexes with CN 6, because their ionic radii are large enough and/or the appropriate nd orbitals are available for hybridization. Examples are $M(H_2O)_6^{3+}$, MF_6^{3-} or $M(acac)_3$; all these species are hypervalent. The triiodides of Al, Ga and In, and the tribromides of Al and Ga, in both the solid and the gas phase form M_2X_6 dimers in which two X atoms act as bridging atoms. Aluminium trichloride is also dimeric, but only in the vapour. Fluorine does not act as a bridging atom because of its high electronegativity, which hinders the sharing of its electrons with two metal atoms. Formation of dimers and adducts with electron pair donors (Lewis bases) can be attributed to the tendency to complete the octet. Aluminium forms stable highly covalent organometallic compounds AlR_3 which, like the trihalides, are Lewis acids.

The small size of the $2p$ orbital, incomplete shielding from the nuclear charge by filled $3d$ and $4f$ shells and relativistic contraction of $6s$ and $6p_{1/2}$ orbitals in Tl make some properties of outer orbitals in Group 13 atoms to oscillate. Examples are s and $p_{1/2}$ orbital radii and energies as well as the ionization potentials I_1, I_2 and I_3 (Figs. 9.3 and 5.3, and Table 9.1). This oscillatory behaviour is, as we already know (Section 5.3), called secondary periodicity. With respect to the 3+ ionic radii and with respect to metallic radii the incomplete shielding from the nuclear charge by filled $3d$ and $4f$ shells divides the elements from Al to Tl into two subgroup pairs: Al, Ga and In, Tl (Fig. 9.1). Since the aqueous chemistry of these elements is that of the M^{3+} cations (M^{3+} and M^+ for Tl), similarity or dissimilarity in 3+ ionic radii makes many chemical and crystallochemical properties of Al^{3+} and Ga^{3+} ions on the one hand and In^{3+} and Tl^{3+} ions on the other similar. For instance, the coordination chemistry of Al^{3+} and Ga^{3+} cations is very similar, differing significantly from that of In and Tl. The hydroxides of both Al and Ga are amphoteric, those of In and Tl do not dissolve in alkalis. Ga_2O_3 has an allotropic modification isostructural with corundum, Al_2O_3, whereas In_2O_3 and Tl_2O_3 both have (at normal pressure) the CaF_2 structure with ¼ of the anions missing. Al and Ga are thus in 6-coordination, In and Tl in 8-coordination, in these sesquioxides. Both Al and Ga, like boron, form hydride anions MH_4^-. Lithium aluminium hydride, $LiAlH_4$, a

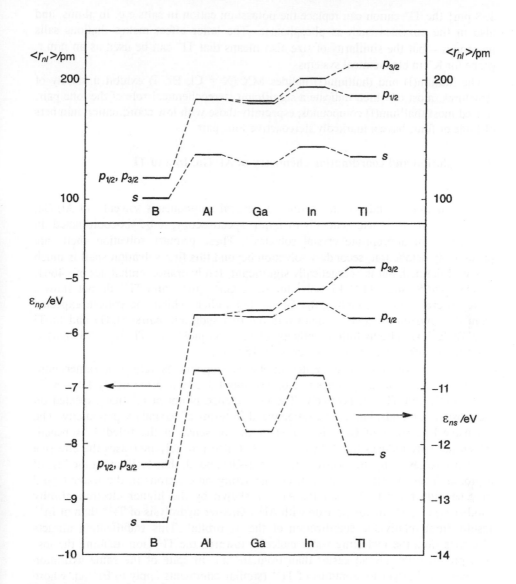

Fig. 9.3 Changes of orbital radii and energies down Group 13.

crystalline solid stable in the dry below about 120 °C, is a powerful reducing agent, important in organic chemistry; sodium borohydride, $NaBH_4$, is a gentler alternative.

Indium and thallium show an accessible and a stable +1 oxidation state respectively. Indeed for thallium the +1 oxidation state is even more stable than the +3. Thus, for example, TlI_3 is thallous tri-iodide, $Tl^I(I_3)$, rather than thallic iodide, $Tl^{III}I_3$. Tl(I) shows some resemblance to Ag(I) in the low solubility of its halide salts. Having the same charge and similar ionic radius {$r_i(Tl^+) = 150$ pm, r_i (K^+) =

138 pm} the Tl^+ cation can replace the potassium cation in salts e.g. in alums, and also in the enzymes such as phosphatase. The latter effect makes thallous salts poisonous – but the similarity of size also means that Tl^+ can be used as an n.m.r. probe for K^+ in biochemical systems.

The indium(I) and thallium(I) halides MX (X = Cl, Br, I) exhibit a variety of structures, most of which indicate a significant stereochemical role of the lone pair. Indeed most thallium(I) compounds, especially those with low coordination numbers of three or four, have a markedly stereoactive lone pair.

9.3.2 Solution and coordination chemistry of Al, Ga, In and Tl

Ion solvation

The 3+ aqua-ions, and solvated cations in several non-aqueous solvents, of Al, Ga, and In have been shown, generally by n.m.r. spectroscopy, to be six-coordinated, in solution as in appropriate crystal solvates. These primary solvation shells are presumably octahedral; secondary solvation beyond this first solvation shell is much less well-defined, but is energetically significant. Ion hydration enthalpies of -4661, -4685, -4108, and -4184 kJ mol^{-1} for Al^{3+}, Ga^{3+}, In^{3+}, and Tl^{3+} do not show a steady decrease on descending the Group, but rather exhibit the pattern expected from the division of $r_i(M^{3+})$ values into the two subgroup pairs Al, Ga and In, Tl (see Table 5.4). The hydration enthalpy of -330 kJ mol^{-1} for Tl^+ ($r_i = 150$ pm) is close to that of -299 kJ mol^{-1} for Rb^+ ($r_i = 152$ pm).

Polarization of coordinated water in the M^{3+}aq ions is reflected, rather non-quantitatively, in the pK_a values of 5.0, 2.6, about 3, and approximately 1 for Al^{3+}, Ga^{3+}, In^{3+}, and Tl^{3+} respectively. The dependence is not at all that expected on electrostatic grounds, but can be rationalized in terms of secondary periodicity. The extensive hydrolysis of Ga^{3+} is caused by the presence of the filled $3d$ subshell, which not only reduces the radius of Ga^{3+} but also probably increases the effective nuclear charge felt by the electrons in the $Ga–OH_2$ bond, thus facilitating the loss of a proton. That the effective nuclear charge acting on electrons in the water ligand may be higher for Ga^{3+} than for Al^{3+} is shown by the higher electron affinity (higher I_3) of Ga^{3+} in comparison with Al^{3+}. Greater hydrolysis of Tl^{3+} than of In^{3+} results from relativistic stabilization of the $6s$ orbital. This stabilization attracts electrons from the hydrating water molecule towards the Tl^{3+} ion, making the loss of a proton from Tl^{3+}aq easier than from In^{3+}aq, in spite of the same hydration number 6 and larger ionic radius of Tl^{3+} (similar comments apply to Bi^{3+}aq, whose pK_a is also unexpectedly low, ~ 1.8, on electrostatic grounds; see also Section 11.2). It may be added that strong electron attraction by other oxidizing metal ions gives rise to similarly enhanced acidity, as for example for Co^{3+}aq ($pK_a \sim 1$) and Mn^{3+}aq ($pK_a \sim 0$).

Complex stabilities

Stability constants for halide complexes of the Group 13 metal cations (Table 9.2) provide a good illustration of the Chatt and Duncanson "Class a"/"Class b" and Pearson "Hard and Soft Acids and Bases" (HSAB) generalizations. The most stable combinations are "hard"/"hard" in AlF^{2+} and "soft"/"soft" in TlI^{2+}. In accordance with the Chatt-Duncanson classification, stability constants for halide complexes of very "hard" Al^{3+} and significantly less "hard" Ga^{3+} decrease in the order

$$F^- \gg Cl^-, Br^-, I^-$$

whereas for "soft" Tl^{3+} the order is reversed

$$F^- < Cl^- < Br^- < I^-.$$

Complementarily, the stability order

$$Al^{3+} > Ga^{3+} > In^{3+}$$

applies to the series of complexes of the "hard" ligand F^-, whereas the reverse order

$$Tl^{3+} > In^{3+} > Ga^{3+} > Al^{3+}$$

applies to the "soft" ligands I^-, Br^- and Cl^-.

Table 9.2 – Stability constants $(\log_{10} K_1)^a$ for complex formation between Group 13 M^{3+}aq cations and halide ligands

	Al^{3+}	Ga^{3+}	In^{3+}	Tl^{3+}
fluoride	6.1	4.4	3.8	b
chloride	b	0.1	2.3	6.7
bromide	b	~ 0	2.0	9.0
iodide	b	~ 0	~ 2	36

a Comparisons should be made with some caution, because the accuracies of some of these $\log_{10} K$ values may be no better than ± 1.

b Values too small to measure accurately.

Solvent exchange and complex formation

Mechanisms of solvent exchange and complex formation provide another interesting and important illustration of the subdivision of the Group 13 elements into the two pairs Al, Ga and In, Tl. The combination of known solvation geometries and moderate rates of solvent exchange at the Group 13 M^{3+} cations, at least for M = Al, Ga and In means that both can generally be investigated by n.m.r. techniques. This permits establishment of kinetic parameters and assignment of mechanism. The variation in activation entropies ($\Delta S^{\ddagger}$) and volumes ($\Delta V^{\ddagger}$) for solvent exchange going down the Group (Table 9.3) indicates the assignment of dissociative mechanisms to solvent exchange at Al^{3+} and Ga^{3+}, but associative exchange at In^{3+}.

Rate laws for trimethyl phosphate (TMP) and for dimethyl sulphoxide (DMSO) exchange at Al^{3+} and at Ga^{3+} in nitromethane as non-coordinating cosolvent are zero-order in TMP or DMSO, again suggesting dissociative exchange at these centres. On the other hand, TMP exchange at In^{3+} in nitromethane is first-order in TMP, consistent with associative exchange. Thus activation parameters and rate laws give consistent mechanistic indications, which are in accordance with the intuitive expectation of associative mechanisms being easier for larger cations, where there is more room to accommodate the incoming group in forming the transition state. The division into two groups, Al and Ga *vs* In and presumably Tl, is thus confirmed.

Table 9.3 – Activation entropies and volumes for solvent exchange at Group 13 solvento-cations.

	water		dimethyl sulphoxide		trimethyl phosphate	
	$\Delta S^{\ddagger}$	$\Delta V^{\ddagger}$	$\Delta S^{\ddagger}$	$\Delta V^{\ddagger}$	$\Delta S^{\ddagger}$	$\Delta V^{\ddagger}$
Al^{3+}	+42	+6	+22	+16	+37	+23
Ga^{3+}	+30	+5	+4	+13	+27	+21
In^{3+}	−96				−113	−21

Complex formation at Al^{3+} and at Ga^{3+} is also dissociative in character, from rate law and activation parameter evidence. Complex formation from aluminium(III) is still dissociative in character at higher pHs where the main reactant is $AlOH^{2+}$aq.

9.4 GROUP 3: Sc, Y, and La

Table 9.4 — Fundamental properties of the Group 3 elements, and of aluminium and lutetium for comparison

	Al	Sc	Y	La	→	Lu
$R = \langle r_{out} \rangle$ / pm	181	208	223	250	→	206
I_1/eV	5.98	6.54	6.38	5.58	→	5.42
$I_2 + I_3$/eV	47.3	37.6	32.8	30.2	→	34.8
χ	1.61	1.36	1.22	1.10	→	1.27
mp (°C)	660	1541	1522	921	→	1663
r_{met} / pm	143	161	181	188	→	173
$r_i(3+)$ / pm (CN 6)	53.5	74.5	90	103.2	→	86.1
$E°(3+/0)$ / V	−1.68	−2.03	−2.37	−2.38	→	−2.30

Table 9.4 shows some values for fundamental properties of these elements, together with the respective values for aluminium and lutetium for comparison. It should be noted that lutetium, commonly classified as a lanthanide, is in fact a *d* electron element, because it has a filled 4*f* subshell with the electron configuration $4f^{14}5d^{1}6s^{2}$ (see Section 18.1). The Group 3 and 13 elements have not very much in

common, although previously they were labelled as Groups IIIB and IIIA elements. The main property in common is a maximum oxidation number of +3. However, in respect to oxidation states the Group 3 elements are more similar to aluminium than to the lower Group 13 elements Ga, In and Tl. Thus Sc, Y and La, like Al, do not show the oxidation number +1 for reasons explained for the example of magnesium which also has two electrons in the outermost s orbital (see Section 5.2). That aluminium, in spite of its s^2p^1 electron configuration, i.e. with one electron in the outermost subshell, does not show the oxidation number +1 is the result of a combination of the highly negative hydration enthalpy of the Al^{3+} ion and the moderate sum of the ionization potentials I_2+I_3 (see Section 5.3).

Comparison of the data in Tables 9.1 and 9.4 shows that I_1 is similar for elements in both Groups (except for boron), whereas the sum of the ionization potentials I_2+I_3 is markedly lower for Group 3 than Group 13 elements. That I_1 is similar, in spite of the significantly greater radius of the outermost orbital in Group 3 than in Group 13 elements, is probably the result of the greater tendency of the s than of the p electron to penetrate the atomic core, where the electron is much less screened from the nuclear charge. On the other hand the sum of ionization potentials I_2+I_3 is greater for Group 13 than for Group 3 elements (in the same Period) because of the greater effective nuclear charge which, in turn, results from the presence of the filled d subshell. Another reason is that in the Group 3 elements the third electron is removed from the not-too-distant $(n-1)d$ subshell (see Section 15.3).

The lower effective nuclear charge acting on the outermost electrons in the M^{3+} ions of Group 3 is also the reason why the ionic radii of elements in the same Period are higher by 10 to 14 pm for Group 3 than for Group 13 elements. Because of their greater radii Sc^{3+}, Y^{3+}, and La^{3+} interact more reluctantly than Group 13 M^{3+} ions with simple ligands, forming complexes of only moderate stability. Another factor which disfavours complex formation by Group 3 tripositive ions is their lower polarizability which, as we know from Section 4.5, is related to hardness (or rather to softness). For instance hardness, as calculated from eq. 4.11, is 24.4 for Sc^{3+} and only 16.8 eV for Ga^{3+}. This means much lower polarizability for Sc^{3+} and its reluctance to form complexes with partially covalent bonding. The two Groups differ considerably also in respect to coordination number of the M^{3+} ions in complexes. Whereas in Group 13 the highest coordination number is almost always 6, it increases from essentially 6 for scandium to 9 or even higher for lanthanum.

pK_a values for M^{3+}aq increase steadily from M = Al (pK_a = 5.0 – acidity comparable to acetic acid) to M = La (pK_a ~ 9), as expected for increasing ease of loss of a proton from coordinated water under the influence of the adjacent M^{3+} ion on going from La to Al. In fact trends in properties of the aqua-ions of Group 3, plus Al^{3+}aq, although predominantly determined by electrostatics (i.e. size), are not entirely regular. They show some reflection of the change in hydration numbers from 6 for Al^{3+}, through perhaps 7 for Sc^{3+} and 8 for Y^{3+} (as in the crystal hydrate $YTcCl_6.8H_2O \equiv [Y(H_2O)_8][TcCl_6]$), to 9 for La^{3+}. This variable pattern contrasts with solvation numbers of six in all but the bulkiest solvents for Al^{3+}, Ga^{3+} and In^{3+}. The greater coordination numbers of Y^{3+} and La^{3+}, and larger ionic radii, contribute to the lesser degree of hydrolysis of Group 3 aqua-cations compared with their Group 13 analogues. Rate constants for water exchange at the Group 3 aqua-cations increase from ~ 1 s^{-1} for Al^{3+}aq to ~ 5 x 10^8 s^{-1} for La^{3+}aq. Reactivities for both Group 3 and Group 13 aqua-cations correlate well with reciprocal ionic radii.

Somewhat unexpectedly, the melting points (Tables 9.1 and 9.4) and boiling points of the Group 3 elements are much higher than those of the Group 13 elements Al to Tl (though boron has exceptionally high melting and boiling points, even higher than the Group 3 elements). The reason is the much smaller sum of the ionization potentials $I_1 + I_2 + I_3$ for Group 3 than for Group 13 elements. Why ionization potentials affect the atomization energy and thus melting and boiling points of metals can easily be explained. To this end one can suppose that a metal is formed in two imaginary steps. In the first, the separated atoms are ionized to the extent that the valence electrons are detached. In the second step the ionized constituents are brought together to form a lattice of positive ions immersed in a fluid consisting of itinerant electrons. From the Born-Haber cycle based on this model it becomes evident that the atomization energy of a metal is the larger the lower is the expenditure of energy to ionize the atoms. The atomization, or cohesive, energy is defined as the energy required to separate the metal into isolated atoms. Also unexpectedly, the melting point of lanthanum is much lower than the very similar melting points of scandium and yttrium, which is probably the result of the presence of f orbitals in the energy band of lanthanum metal (see Section 18.2). Sc, Y and La show electrical conductivity lower than Ga, In and Tl and much lower than Al, which is the fourth best conductor. With respect to conductivity yttrium is of considerable interest and importance as an essential component of a number of superconductors, such as the relatively long-known perovskite-related $YBa_2Cu_3O_7$, $YSr_2Cu_2GaO_7$, and carbide halides $Y_2C_2X_2$ (X = Cl, Br, I). Superconductivity in the $Y_2C_2X_2$ systems is connected with π-back-bonding between p orbitals of the C_2 units and vacant d orbitals on the yttrium.

10

Group 14

Table 10.1 – Fundamental properties

	C	Si	Ge	Sn	Pb
$R = \langle r_{np} \rangle$ / pm	92	147	152	168	163
$I_1 + I_2$/eV	35.6	24.5	23.8	22.0	22.4
$I_3 + I_4$/eV	112.3	78.6	79.9	71.2	74.2
χ	2.55	1.90	2.01	1.96	2.33
mp (°C)	~3540 [a]	1410	938	232	328
r_{cov} / pm	77.2	117	122	141	150
r_{met} / pm			139	158	175
r_i(2+) / pm (CN 6)			73	93	119
r_i(4+) / pm (CN 6)		40.0	53.0	69.0	77.9
$E°$(2+/0) / V				−0.14	−0.125
$E°$(4+/2+) / V				+0.15	+1.69

[a] Diamond

As in Group 13 the first element in Group 14 – carbon – differs very much from its heavier congeners with respect to both fundamental properties of the atom and chemical properties of the element. The unique properties of carbon result from its high ionization potentials, small atomic radius, high electronegativity, and unavailability of d orbitals (see Table 10.1 and Fig. 10.1). The higher ionization potentials and smaller atomic radius are, in turn, caused by the unique properties of the $2p$ shell (Fig. 10.2; see Section 3.2). Because of the very small radial extent of both the $2p$ and $2s$ orbitals carbon, like boron, has a much higher atomization enthalpy and melting temperature than its heavier congeners silicon and germanium.

10.1 PROPERTIES OF CARBON

For reasons explained in Section 6.5 carbon does not form a metallic phase. Instead it forms a variety of covalent solid phases of which the most important are diamond and graphite. The band structure of diamond has already been discussed. In graphite carbon is sp^2 hybridized and the lower band is filled because it contains

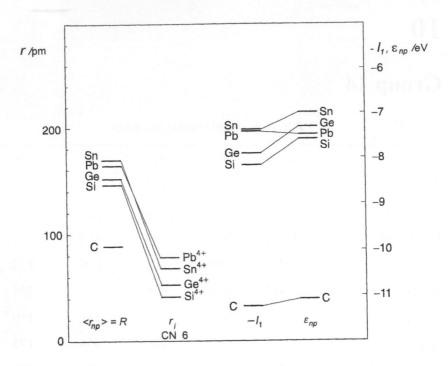

Fig. 10.1 Radii, energy of np orbitals and first ionization potentials of Group 14 elements.

only bonding sp^2 orbitals ($k = 3$, $z = 3$). The remaining perpendicular p_z orbitals, one on each carbon atom, overlap to form π bonds and then with increasing number of carbon atoms delocalize over the plane forming the upper band which is just in touch with the lower band. Since the p band consists of both bonding and antibonding $p\pi$ orbitals it is half-filled. Therefore, graphite is a metallic conductor parallel to the planes and a semiconductor in the perpendicular direction. Since in graphite the $p\pi$ band is not filled, it can accept electrons, which means that graphite can be reduced by e.g. potassium to form the intercalation compound KC_8 in which the K^+ cations are inserted between the graphite sheets. Also fullerenes, which are related to graphite (see Section 6.1), can be reduced to form e.g. K_3C_{60}.

In most carbon compounds, even with highly electronegative Group 17 elements, there is a large covalent contribution to bonding. Carbon is as a rule tetravalent and shows the formal oxidation number +4. Carbon shows a formal oxidation number of +2 only in the CO molecule, and is divalent only in carbenes. Carbenes CR_2 with R = alkyl are very unstable, CX_2 where X = halide a little less unstable. However carbenes CR^1R^2 with half-lives of several weeks can be synthesised by appropriate choice of R^1 and R^2, such relatively high stability being achieved by balancing mesomeric and inductive effects, e.g. by having $R^1 = -C_6H_3(CF_3)_2$ and $R^2 = -P(N^1Pr_2)_2$.

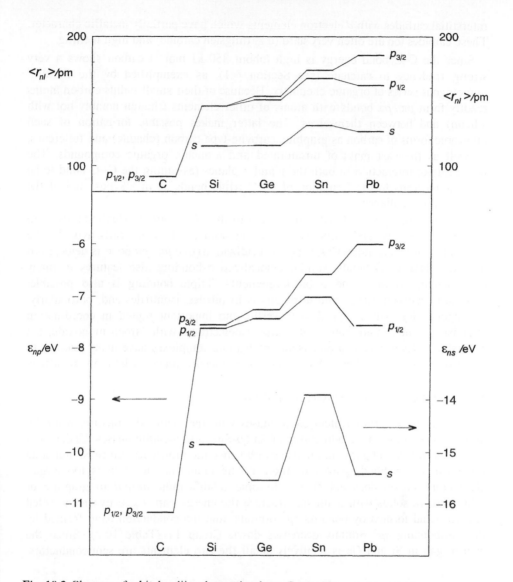

Fig. 10.2 Changes of orbital radii and energies down Group 14.

Because of its relatively high electronegativity carbon shows formal negative oxidation numbers in compounds with electropositive elements of Groups 1, 2 and 13. These are so-called salt-like carbides e.g. CaC_2 which contains the acetylide anion $C{\equiv}C^{2-}$ or Al_4C_3 which contains the C^{4-} anion. The formal oxidation number of carbon is -1 in the former, -4 in the latter. In reaction with water CaC_2 yields acetylene (ethyne), while Al_4C_3 yields methane. With silicon and boron, which are not much less electronegative, carbon forms covalent carbides as e.g. SiC (carborundum) and B_4C. The latter is even harder than diamond. Carbon forms interstitial carbides with d electron elements which have partially metallic character. These carbides too are often very hard (e.g. tungsten carbide) and high melting.

Since the C–C bond energy is high (about 350 kJ mol^{-1}) carbon shows a very strong tendency to catenate (see Section 6.1), as exemplified by the familiar homologous series of organic chemistry. Because of their small radius carbon atoms readily form $p\pi$–$p\pi$ bonds with atoms of other elements (though notably not with silicon) and between themselves. The latter makes possible formation of such allotropic forms of carbon as graphite, carbyne-type carbon (chaoite) and fullerenes, as well as the vast mass of unsaturated and aromatic organic compounds. The possibility of interaction in both the y and z planes (assuming the C–C bond to be the x axis) leads to the formation of C≡C triple bonds as in alkynes, and of the C=C=C unit of allenes.

Carbon forms a $p\pi$–$p\pi$ bond with oxygen in the >C=O group which is present in aldehydes, ketones, and carboxylic acids, as well as in metal carbonyls. In the triangular carbonate anion $CO_3{}^{2-}$ (sp^2 hybridization) one $p\pi$–$p\pi$ bond is delocalized over the three C–O bonds; similar delocalized π-bonding also features in many other oxoanions of p and d block elements. Triple bonding is also possible, expecially between carbon and nitrogen, as in nitriles, isonitriles and, particularly, the cyanide ion which is such a powerful and important ligand in coordination chemistry. The cyanide ion is, of course, isoelectronic with carbon monoxide; the extensive series of metal carbonyls and of cyanide complexes have many features in common and provide a key link between organometallic and coordination chemistry.

10.2 PROPERTIES OF Si, Ge, Sn and Pb

Si, Ge and α-Sn form covalent solid phases with the diamond structure, while Pb and the other allotropic modification of Sn (β-Sn) form metallic phases. Under high pressures silicon and germanium also exhibit metallic properties. Since Si atoms do not form $p\pi$–$p\pi$ bonds with one another (the radius of the atom is too large), elementary silicon does not show allotropic modification similar to graphite or fullerenes. In solids with diamond structure the energy gap, E_g, between the filled valence band formed by bonding sp^3 orbitals, and the conduction band formed by the antibonding sp^3 orbitals decreases down Group 14 (Table 10.2). Since the energy gap in Si and Ge is relatively small the two elements are semiconductors.

Table 10.2 – The energy gap in Group 14 elements

Element	C	Si	Ge	α-Sn	Pb
E_g/eV	~ 6.0	1.10	0.67	0.08	0.0
$E_g/kJ\ mol^{-1}$ [a]	~ 600	106	64	7.8	0

[a] Band gaps for Group 13/15 (III/V) semiconductors range from 218 kJ mol^{-1} for GaP down to 17 kJ mol^{-1} for InSb.

The increase in metallic character down the p block Groups has been explained in Section 6.5. It is interesting to note (see Table 10.1) that the metallic radius is much larger than the covalent radius. The reason is that metallic bonding, contrary to covalent bonding, is electron-deficient. Hence it is weaker and the corresponding radius significantly larger.

As in the case of Group 13 elements, the presence of filled $3d$ and $4f$ shells and the direct relativistic effect generate secondary periodicity in such properties of atoms as for example ε_{ns}, $\langle r_{ns} \rangle$, and the summed ionization potentials $I_3 + I_4$ (Figs. 10.2 and 5.4, and Table 10.1). Interestingly, the sums of the first and second ionization potentials for Sn and Pb ($I_1 + I_2 = 22.0$, 22.4 eV, respectively) are very close to that for Mg ($I_1 + I_2 = 22.6$ eV). Filled $3d$ and $4f$ shells divide the elements from Si to Pb into two subgroups: Si, Ge and Sn, Pb. With respect to a number of fundamental atomic properties the presence of the filled $3d$ shell brings Ge closer to Si, while the filled $4f$ shell and relativistic effects shift Pb towards Sn. The elements in each subgroup show similar atomic and ionic (4+) radii, and also similar values for the summed ionization potentials $I_1 + I_2$ and $I_3 + I_4$ (Fig. 10.1 and Table 10.1). Changes in the energy gap E_g (Table 10.2) provide another example of the division into Si, Ge and Sn, Pb subgroups. The particularly large energy gap between Si and C is consistent with the unique properties of each first element in a p block Group. The pairs of elements in the Si, Ge and Sn, Pb subgroups are also more alike with respect to chemical properties, e.g.:

– germanium dioxide is known in structures analogous to those of SiO_2;

– both silicon and germanium are semiconductors, whereas white (β) tin and lead are metals;

– there is a marked decrease in stability of the +4 oxidation state between the two subgroups.

The elements from Si to Pb form both covalent and ionic bonds. Ionic bonding is of importance in complexes with highly electronegative anions, e.g., in SiF_6^{2-}, GeF_6^{2-}, SnF_6^{2-} and $PbCl_6^{2-}$. The formation of six-coordinate complexes, which are hypervalent compounds, can be explained by participation of nd orbitals in hybridization (sp^3d^2 hybridization), although formation of these complexes can also be explained without referring to such nd orbital participation (see Section 5.1).

Covalent bonding is typical for analogues of hydrocarbons. As pointed out in Section 6.1, Si, Ge, Sn and Pb, in contrast to carbon, form only short chain hydrides M_nH_{2n+2} where $n \leq 6$ for Si, $n \leq 9$ for Ge, $n \leq 2$ for Sn and $n = 1$ for Pb – a further example of differences between the Si, Ge and Sn, Pb subgroups. The low tendency to catenate, even for silicon, is the result of the much smaller energy of the Si–Si than the C–C bond. However, the first three elements when bonded to two fluorine atoms or two alkyl groups show a stronger tendency to self-link. The structural unit in the long chains which are formed is the $-MR_2-$ group. Formation of compounds such as $Sn(SnPh_3)_4$, clusters of the cubane type, $(MR)_8$, and of anionic clusters (see below, and Section 6.2) by Ge, Sn and Pb is another manifestation of the tendency for self-linking.

Silanes Si_nH_{2n+2} are much more reactive than alkanes with respect to hydrolysis (contrary to methane, monosilane reacts violently with water). The main reason is the high Si–O bond energy, which is considerably higher than that of Si–Si and Si–H bonds. Also kinetic factors contribute to the lower stability of silanes than hydrocarbons. Fast hydrolysis of silanes can be explained assuming availability of $3d$ orbitals in the Si atom. Participation of a $3d$ orbital in hybridization facilitates formation of a transition state containing the incoming OH^- ion (the CN of the Si atom is then equal to 5), which accelerates hydrolysis. The Si atoms form $p\pi-p\pi$ bonds only in exceptional cases, but easily form $p\pi-d\pi$ bonds with atoms of the $2p$-row. In such bonds the empty $3d$ orbitals of the silicon act as electron acceptors. The spatial characteristics of d orbitals do not require very small radii for both atoms for the formation of $p\pi-d\pi$ bonds. Thus $p\pi-d\pi$ bonding is present in siloxanes that contain alternating Si and O atoms as e.g. in the cyclic dimethylsiloxane $[(Me_2SiO)_n]$, making the siloxane chain more stable than its silane analogue.

A characteristic property of silicon is formation of silicates, whose basic unit is the tetrahedral SiO_4^{4-} anion. By sharing 1, 2, 3 or 4 oxygen atoms SiO_4^{4-} tetrahedra form dimeric anions $Si_2O_7^{6-}$ (one oxygen atom in common), cyclic anions with the empirical formula $Si_3O_9^{6-}$ and $Si_6O_{18}^{12-}$ (two atoms), infinite chain anions $(SiO_3^{2-})_n$ (two atoms), infinite sheet anions $(Si_2O_5^{2-})_n$ (three atoms) and $(SiO_2)_n$ (4 atoms in common). Aluminosilicates are formed when aluminum atoms replace some silicon atoms in the framework of SiO_4^{4-} tetrahedra. The lattice then becomes negatively charged and cations, mainly from Group 1 or 2, must enter the lattice to neutralize the charge.

In accordance with the general rule concerning oxidation numbers in p block Groups (see Section 5.2) the elements from silicon to lead display even oxidation numbers +4 and +2. The stability of the lower oxidation state increases down the Group, just as in Group 13. Thus, SiF_2 and SiO are unstable, while germanium displays the oxidation state +2 only with halogens (in compounds GeX_2 and in the complex $[GeF_3]^-$) and in the chalcogenides GeS, GeSe, and GeTe, with the preparation of GeS and GeSe requiring strong reductants. Tin and lead, however, form stable oxides SnO and PbO, sulphides SnS and PbS, and dihalides, while hydrated Sn^{2+} and Pb^{2+} cations occur in crystal hydrates and in aqueous solutions of tin(II) and lead(II) salts. Tin(II) salts are mild reductants, but lead(IV) compounds (lead dioxide, lead tetraacetate) are strong oxidants. The remarkable stability of the Pb^{2+} ion, like that of Tl^+, is caused by the direct relativistic effect and by the presence of the filled $4f$ subshell, which stabilize the $6s$ electrons. There are a number of mixed valence lead comounds, of which the best known is red lead, Pb_3O_4. This is a valence-localized compound (Class I in the Robin and Day classification) – the lead(II) and lead(IV) centres are crystallographically distinct. However there is sufficient charge-transfer to endow Pb_3O_4 with its distinctive bright red colour, for which it has been valued as a pigment for centuries.

Ge, Sn and Pb on reduction by Na in liquid ammonia form anionic clusters $[M_5]^{2-}$ and $[M_9]^{4-}$ in which the formal oxidation number is -0.4 and -0.44, respectively (cf. -1 in the alloy NaPb). These formal oxidation states, unusual for Group 14 elements, are stabilized by metal-to-metal bonds in the cluster and by formation of salts with large cations, such as the red crystalline $[Na(crypt)]_2[Sn_5]$, where crypt = the cryptand [222] (whose formula is shown in Fig. 7.5 on p. 75). The bond orders in the clusters are low. For instance, in the $[M_5]^{2-}$ clusters, which have (distorted) trigonal bipyramidal structures, there are 5 nonbonding electron pairs (one on each metal atom), which leaves 12 electrons for 9 bonds. The Sn–Sn distances in $[Sn_5]^{2-}$ are in the range 285 to 310 pm, which lies between $2r_{cov}(Sn) = 282$ and $2r_{met}(Sn) = 316$ pm; the Sn–Sn single bond distance in Sn_2Ph_6 is 277 pm.

The $5s^2$ and $6s^2$ electron pairs of Sn and Pb, respectively, are often stereochemically active, although pairs of tin(II) and lead(II) compounds are rarely isostructural. Stereochemical effects of these ns^2 lone pairs tends to be more evident for tin(II) than for lead(II), and more marked for electronegative ligands. The structural consequences range from negligible for PbS, which has the NaCl structure, to species with very stereoactive lone pairs, such as pyramidal $SnCl_3^-$ in its Cs^+ salt, and SnF_2 and its adduct $SnF_2.AsF_5$. This adduct is a trimer which can to a first approximation be regarded as ionic, $[Sn_3F_3][AsF_6]_3$, containing cyclic $[Sn_3F_3]^{3+}$ cations. But the AsF_6^- anions, as so often in Lewis acid-base adducts of this type (cf. $XeF_2.SbF_5$ in Section 14.2), form strong $Sn\cdots F$–As bridges. The four fluorines closest to each tin (its two neighbours in the Sn_3F_3 ring, at 210 pm, and two fluorides forming the bridges to adjacent AsF_6^- anions, at 259 pm) all lie on the same side of the plane normal to the tin–lone pair axis (Fig. 10.3 – where F_c represents a fluoride in the $[Sn_3F_3]^{3+}$ ring, F_{as} a bridging fluoride from AsF_6^-, and lp the lone pair on the tin).

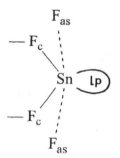

F_{as}

— F_c

Sn   lp

— F_c

F_{as}

Fig. 10.3 Projection of an $Sn(F_c)_2(F_{as})_2(lp)$ unit (see text) of $[Sn_3F_3][AsF_6]_3$ onto a plane perpendicular to the base of the approximately square-pyramidal $Sn(F_c)_2(F_{as})_2(lp)$ moiety.

Whereas the stereochemistries of silicon and of germanium are generally straightforward, with the great majority of compounds tetrahedral and a small number of octahedral complexes, the stereochemistries of lead and, especially, of tin are considerably more varied. In part this arises from the latter pair of elements having two stable oxidation states, and in turn from the variable stereochemical effect of the ns^2 electron pairs in compounds of the lower oxidation state. Tin(IV) occurs quite often in five or six coordination, as well as in the typical tetrahedral stereochemistry; the stereochemistry of tin(II) is complicated not only by the presence of the lone pair but also by the marked tendency of tin(II) to increase its coordination number by forming rings (cf. $[Sn_3F_3]^{3+}$ above) and chains. Lead exhibits a wide range of coordination numbers. The lead(II) in Pb_3O_4 is 3-coordinate; lead(IV) is often 4-coordinate, as for example in most organolead compounds. Octahedral lead(IV) occurs in complexes of the $[PbCl_6]^{2-}$ type, and in the form of octahedral PbO_6 units in Pb_2O_3, Pb_3O_4, one form of PbO_2, and in mixed oxides such as Sr_2PbO_4. In keeping with its large size, lead sometimes has a high coordination number, seven in the alloy NaPb, eight in several compounds and complexes, including PbF_2 (fluorite structure), PbFCl, and lead tetra-acetate, and nine (albeit as seven near and two somewhat less near atoms) in $PbCl_2$. The structure of PbFCl is particularly interesting from the point of view of the relation between radius ratios and packing. Isolated MX_8 units are generally more stable in square antiprismatic than in cubic form. However the usual structure for salts MX with 8:8 coordination is the cubic CsCl structure – it is easy to pack cubes to form an infinite three-dimensional array, but to pack square antiprisms is awkward and an inefficient use of space. But if the antiprisms each have one small square face and one large square face, preferably with the diagonal of one square equal in length to the side of the other, then they may be packed efficiently into a three-dimensional array. It so happens that the pairs of anions F^- and Cl^-, and O^{2-} and S^{2-}, conform quite well with the implied radius ratio of $1/\sqrt{2}$, so with a sufficiently large central cation square antiprismatic MX_4Y_4 units can be efficiently packed into an infinite three-dimensional layer structure, to give what is known as the lead chloride fluoride structure. This is outlined schematically in Fig. 10.4; this structure is also adopted by, e.g., BiOCl and UOS.

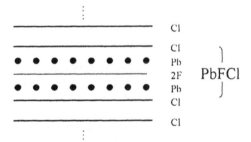

Fig. 10.4 Outline of the PbFCl structure

10.3 COMPARISON WITH GROUP 4 ELEMENTS

Although our main discussion of the transition elements comes later in this book, it is convenient here, as in preceding and following Chapters, to establish comparisons between related pairs of p- and d-block Groups. Table 10.3 lists some properties of the Group 4 elements for comparison with their Group 14 analogues (Table 10.1).

In fact, except for the maximum oxidation number +4, the Group 4 elements have little in common with the Group 14 elements, even with the three heaviest (Ge, Sn and Pb). In contrast to tin and lead, zirconium and hafnium do not show a stable +2 oxidation state in simple salts. The reason is the much lower sum of the ionization potentials $I_3 + I_4$ for Zr and Hf, compared with Sn and Pb, which favours detachment of the third and fourth electrons (compare data in Table 10.1 and 10.3). However, Zr (as also Hf) shows a relatively stable oxidation number +2 in the so-

Table 10.3 – Fundamental properties of the Group 4 elements

	Ti	Zr	Hf
$R = \langle r_{ns} \rangle$ / pm	198	210	195
$I_1 + I_2$ / eV	20.4	20.0	21.6
$I_3 + I_4$ / eV	70.7	57.3	56.6
χ	1.54	1.33	1.3
mp (°C)	1660	1852	2230
r_{met} / pm	144.8	160	156.4
r_i(4+) / pm (CN 6)	60.5	72	71

called centred clusters which contain the $[(Zr_6Z)]X_{12}$ unit. This unit consists of an octahedron of Zr atoms with bridging halogen atoms over each triangular face. The interstitial central atom, Z, can be any of a variety of p block or transition elements. The +2 oxidation state, uncommon in Group 4, is stabilized by metal-to-metal bonds in the M_6 cluster. The tendency for self-linking seems to be greater for Ge, Sn and Pb than for Zr and Hf, because the former elements form the so-called "naked" anionic clusters $[M_5]^{2-}$ and $[M_9]^{4-}$, whereas the latter require the additional presence of an interstitial atom and halogen bridging atoms to stabilize the cluster. The greater tendency for self-linking is probably the result of the much smaller radii of Group 14 than of Group 4 atoms.

All Group 4 elements are metals, whereas in Group 14 only β-Sn and Pb show metallic properties. This is because the ionization potentials of Group 4 elements are much lower than those of Group 14 elements in the same rows. Low ionization potentials favour formation of a metallic phase, because valence electrons in metals are delocalized, which can be regarded as partial ionization of the atoms. In analogy to the difference between the Group 3 and 13 elements, the Group 4 elements also have much higher melting points than Ge, Sn and Pb. The reason is the same, i.e. the much lower sum of the first four ionization potentials, I_1 to I_4, for Zr and for Hf than for their Group 14 metallic counterparts Sn and Pb (compare Section 9.4). However, one should note the difference between the melting points of Ti and Ge – that of Ge, which has the diamond structure, is significantly lower.

The ionic radii of tin(IV) and zirconium(IV) are almost the same, those of lead(IV) and hafnium(IV) similar. In spite of this, the highest coordination number in complexes is 8 for the two heavy Group 4 elements, whereas it is 6 for tin(IV) and only rarely greater than six for lead(IV). The probable reason for this difference is that in the hybridization model for complex formation, in order to form complexes with CN 8 the Group 14 elements must utilise four high lying nd orbitals, whereas the Group 4 elements can use the more accessible $(n-1)d$ orbitals.

The Group 4 M^{4+} ions are significantly harder than their counterparts in Group 14. For instance, the hardness of Zr^{4+} as calculated from eq. 4.11 is 23.6 eV, whereas that of Sn^{4+} is only 15.8 eV. Therefore, Group 4 M^{4+} ions form complexes with polydentate ligands containing oxygen and nitrogen donor atoms more readily than their Group 14 analogues. The Group 4 M^{4+} ions form particularly strong complexes with fluoride, hydroxide, and oxide. Indeed, there is a limited chemistry of the ZrO^{2+} moiety (cf. vanadium(IV), VO^{2+}, and the MO_2^+ and MO_2^{2+} oxocations of actinides(V) and (VI)). Despite the fact that pK_a values (of about zero) for Sn^{4+}aq, Zr^{4+}aq, and Hf^{4+}aq and redox potentials involving these aqua-cations have been published and appear in several reference books, there is no definitive evidence for their existence, either in solution or in crystal hydrates. It is likely that in aqueous solution Sn^{4+}, Zr^{4+}, and Hf^{4+} exist in various hydrolyzed and polynuclear forms, depending on pH. However there are several dozen compounds and complexes of Sn^{4+} which contain one or more molecules of water bonded to tin, while the tetranuclear Zr^{4+} species $[Zr_4(OH)_8(H_2O)_{16}]^{8+}$, with four water molecules coordinated to each Zr^{4+}, has been characterized in the solid state (in $ZrOCl_2.8H_2O$) and in solution. The mean Sn^{IV}–O distance to coordinated water is approximately 214 pm, whereas Zr^{IV}–O in the $[Zr_4(OH)_8(H_2O)_{16}]^{8+}$ cation is 227 pm. The longer M–O distance in the latter may be ascribed to the difference in coordination number, 8 for zirconium(IV), 6 in the aqua-tin(IV) complexes, and the greater hardness of Zr^{4+}.

Lead(IV) hydrolyses so easily that under no circumstances does it form a simple aqua-ion, though the simple hydroxo-complex $[Pb(OH)_6]^{2-}$, like $[Sn(OH)_6]^{2-}$, is believed to exist in alkaline solution. It is interesting that there seem to be no analogous $[M(OH)_n]^{(n-4)-}$ species characterised for the harder Zr^{4+} and Hf^{4+} – presumably hydroxo-species of these ions prefer to exist in hydroxo-bridged polynuclear forms. The reason for the ready hydrolysis of lead(IV) is, as in the case of thallium(III), the relativistic stabilization of the $6s$ orbital, which results in attraction of electrons from the water molecules in the aqua-ion toward the Pb^{4+} ion, facilitating in this way the loss of a proton.

11

Group 15

Table 11.1 – Fundamental properties

	N	P	As	Sb	Bi
$R = \langle r_{np} \rangle$ / pm	76.5	125	136	157	168
$I_1 + I_2 + I_3$ /eV	91.6	60.4	56.8	52.5	49.5
$I_4 + I_5$ /eV	175.3	116.2	112.7	100.1	101.3
A /eV	−0.1	0.75	0.81	1.05	0.95
χ	3.04	2.19	2.18	2.05	2.02
bp (°C)	−196	280[a]	616[b]	~1635	~1560
mp (°C)	−210	44[a]	817[c]	631	271
r_i(3+) / pm (CN 6)			58	76	103

[a] White phosphorus (P_4).   [b] Sublimes.   [c] At 3.7 MPa.

Table 11.1 and Fig. 11.1 show that with respect to many fundamental properties of the atom, nitrogen is unique among the Group 15 elements, differing considerably from its heavier congeners. The different chemical properties of nitrogen are the result of high ionization energies, small atomic radius, high electronegativity, non-availability of d orbitals and high bonding energy in the N_2 molecule. Its high ionization energies and small radius are the result of the unique properties of the $2p$ subshell.

11.1 PROPERTIES OF NITROGEN

The stable form of elementary nitrogen, in contrast to the rest of the Group 15 elements, is the dimeric molecule. This is because the N_2 molecule has a very high enthalpy of dissociation, 945 kJ mol^{-1}. Such a high enthalpy of dissociation is the result of the triple bond, which consists of one σ and two π bonds, as depicted in Fig. 11.2. Easy formation of $p\pi$–$p\pi$ bonds is the result of the small size of the nitrogen atom. The much bigger atoms P, As, Sb and Bi do not form $p\pi$–$p\pi$ bonds, hence their stable elementary form is not a dimeric molecule. The N_2 molecule, which is isoelectronic with CO, acts, like the latter, as a ligand in a range of coordination complexes, especially of ruthenium and osmium. Some N_2 complexes, e.g. $[Ru(NH_3)_5(N_2)]^{2+}$, are remarkably stable. Because of the very high bonding energy of the N_2 molecule, fixation of nitrogen (which composes 78% of the earth's atmosphere) is a difficult and energy consuming process. Conversely, but for the same reason, chemical reactions yielding N_2 (burning of gunpowder, decomposition of heavy metal azides) are, as a rule, explosive.

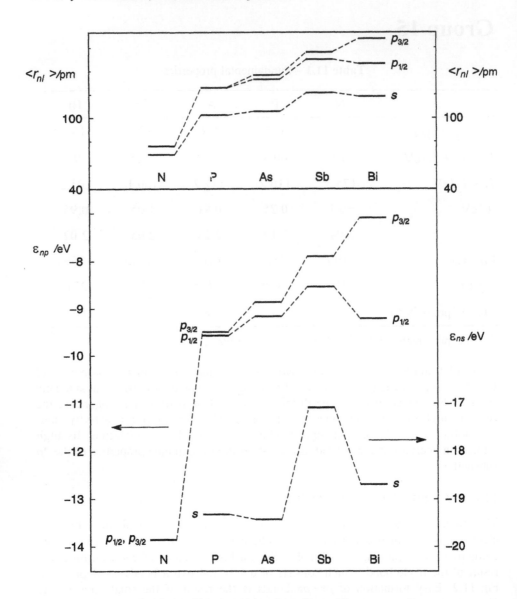

Fig. 11.1 Changes of orbital radii and energies down Group 15.

In spite of its having five valence electrons nitrogen is trivalent and never pentavalent in its compounds with hydrogen and the halogens. The most common explanation is that of the unavailability of the d orbitals which are necessary for

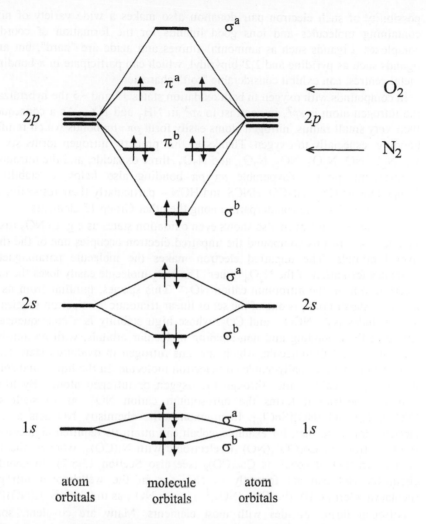

Fig. 11.2 Schematic molecular orbital energy diagram for p block homonuclear diatomic molecule, showing the filling of orbitals in the N_2 and O_2 molecules.

formation of sp^3d hybrid orbitals. However, it has been shown in Section 5.1 that Group 5 elements can be pentavalent, i.e. pentacoordinate, without employing nd orbitals. It seems that the reluctance of nitrogen to be pentacoordinate results mainly from its very small size, which precludes accommodation of five bonding atoms around the central N atom. Nitrogen exhibits all the odd oxidation numbers from −3 to +5. In the NH_3 and NX_3 molecules (X = halogen) nitrogen is in the oxidation states −3 and +3, respectively, and shows sp^3 hybridization with the lone electron pair at one of the corners of the tetrahedron. Due to this lone electron pair the NX_3 compounds are strong electron pair donors and form adducts with electron pair acceptors, as e.g. $R_3N{-}BX_3$ (R = e.g. alkyl; aryl; H) with boron trihalides. The possibility of such electron pair donation also makes a wide variety of nitrogen-containing molecules and ions good ligands for the formation of coordination complexes. Ligands such as ammonia, amines and azide are "hard", but aromatic ligands such as pyridine and 2,2'-bipyridyl, which can participate in π-bonding with metal centres, can exhibit considerable "soft" character.

In compounds with oxygen in both oxidation states +3 and +5 the hybridization of the nitrogen atom is sp^2, in contrast to sp^3 in NH_3 and NX_3. As a consequence of their very small radius, nitrogen atoms easily form $p\pi-p\pi$ bonds which reinforce σ-bonding, especially to oxygen. Thanks to this property nitrogen forms six oxides, viz. N_2O, NO, N_2O_3, NO_2, N_2O_4, and N_2O_5, three oxoacids, and the nitronium and nitrosonium groups. Favourable $p\pi-p\pi$ bonding also helps to stabilize such compounds as HN_3, HNCO, HNCS and HCN – particularly their respective anions. These species have no counterparts among the other Group 15 elements.

Exceptionally, nitrogen also shows even oxidation states as e.g. in NO_2 (oxidation number +4). In this compound the unpaired electron occupies one of the three sp^2 hybrid orbitals. The unpaired electron makes the molecule paramagnetic and promotes formation of the N_2O_4 dimer. The NO_2 molecule easily loses the unpaired electron to form the nitronium cation NO_2^+. This species, familiar from its role in aromatic substitution, is one of the set of linear triatomic 16-electron species, which also includes N_3^-, NCO^-, and CO_2, whose high stability is a consequence of the filling of their bonding and non-bonding molecular orbitals, with no anti-bonding electrons. The NO molecule, which contains nitrogen in oxidation state +2, is the simplest known thermally-stable odd-electron molecule. In the liquid and solid state it dimerizes weakly either through the oxygen or nitrogen atoms. By losing the unpaired electron it forms the nitrosonium cation NO^+, as in such salts as $[NO][SnF_6]$ and $[NO][SbCl_6]$. In organometallic chemistry NO acts as a three-electron donor, so that, for example, cobalt can attain its required eighteen valence shell electrons in $Co(CO)_3(NO)$, isoelectronic with $Ni(CO)_4$, whereas the simplest binary carbonyl of cobalt is $Co_2(CO)_8$ (see also Section 15.5.2). In coordination chemistry NO can act formally as NO^+, as in the well-known nitroprusside (pentacyanoferrate(II), $[Fe(CN)_5(NO)]^{2-}$) ion or NO^-, as in $[Co(NH_3)_5(NO)]^{2+}$.

Nitrogen forms nitrides with most elements. Many are covalent, some are interstitial (very hard and unreactive), and those with electropositive elements of Group 1 and 2, e.g. Li_3N or Mg_3N_2, are, thanks to the high electronegativity of nitrogen, ionic. The ionic nitrides exhibit a range of none-too-simple structures, but it has proved possible to assign a radius of 146 pm to the N^{3-} ion, which is thus slightly larger than O^{2-}.

11.2 PROPERTIES OF P, As, Sb AND Bi

The subdivision of Group 13 and 14 elements into two subgroups (between the fourth and fifth period) is mainly the result of changes in the 3+ and 4+ ionic radii, respectively. However, starting with Group 15 the elements (except for Bi) do not show cationic properties but appear in the form of more or less covalent oxides, oxoacids and halides. Therefore, subdivision into P, As and Sb, Bi subgroups is of much less importance, and one should rather single out phosphorus. The characteristic property of phosphorus is the use of its unfilled d orbitals to form $p\pi-d\pi$ bonds with oxygen and with nitrogen. Because of its smaller radius, phosphorus forms these bonds more easily than silicon. PO_4^{3-} tetrahedra, like SiO_4^{4-}, form chain polyacids $H_{n+2}P_nO_{3n+1}$ (n up to 17) by sharing two oxygen atoms. This is another example of diagonal similarity.

In view of the number of their electrons and orbitals the elements of Group 15 could form typical metallic phases. However, for phosphorus, arsenic and antimony it proves to be energetically more favourable to form tetrahedral M_4 molecules (clusters) which after condensing or polymerizing give various allotropic forms. One such is the extremely reactive white phosphorus, which consists of M_4 molecules bonded by van der Waals forces. White phosphorus is reactive because the P–P bonds formed by the $3p$ orbitals of the P atoms are strongly strained, since the P–P–P angle in the tetrahedral P_4 molecule is only $60°$, whereas the p orbitals are perpendicular to each other. Breaking one of the P–P bonds in each P_4 molecule and linking the fragments into chains or into double layers produces the so-called red and black (metallic) phosphorus, respectively. In the double layer each P atom is bonded to three neighbours by its p orbitals and has one nonbonding s^2 pair. Quasi-metallic layer structures are typical for arsenic and antimony and are the only form for bismuth. The doubly-layered allotropic forms show metallic conduction, the reason for this being relatively close contacts, i.e. interactions, between the double layers. These interactions result in broadening of the bonding, nonbonding and antibonding bands, hence small overlap of the bands leading to metallic character (see Section 6.5). The elements from phosphorus to bismuth show a strong tendency for self-linking with formation of anionic species, such as As_4^{2-} or As_7^{3-}.

Phosphorus, arsenic, antimony and bismuth are tri- and penta-valent and show oxidation numbers +3 and +5. As is usual for p block elements, the stability of the higher oxidation state decreases down the Group. In accordance with this rule bismuth in the +5 oxidation state is a very strong oxidant. However, in contrast to $SbCl_5$, $AsCl_5$ is not known, and arsenates (As^V), are stronger oxidants than antimonates (Sb^V). The variation in oxidation properties from As to Bi is a manifestation of secondary periodicity (see Section 5.3). The much more negative value for the orbital energy ε_{ns} in As than in Sb, which is due to the presence of the filled $3d$ subshell (see Fig. 11.1), makes the sum of the ionization potentials $I_4 + I_5$ for arsenic "abnormally" high (see Table 11.1). This stabilizes the +3 oxidation state for As. One can also say that the highly negative value of ε_{4s} results in high promotion energy of an electron from the $4s$ to a $4d$ orbital. This promotion is necessary for sp^3d hybridization, which according to valence bond theory is shown by the hypothetical $AsCl_5$ and by $SbCl_5$ (see however Section 5.1). The higher promotion energy required for $AsCl_5$ rationalizes its non-existence. The low stability of the +5 oxidation state in Bi results from the presence of the filled $4f$ shell and from the relativistic effect, both of which stabilize the $6s$ electrons. When the electron is removed from the $p_{3/2}$ orbital in the Bi atom, which has the $6s^2 6p_{1/2}^2 6p_{3/2}^1$ configuration, the $p_{1/2}$ orbital becomes the outermost orbital in the Bi^+ cation. This orbital is relativistically stabilized i.e. it has a more negative energy and smaller radius than the $p_{3/2}$ orbital (Fig. 11.1). We know from considerations in

Chapter 5 that smaller radius means greater stability of the corresponding oxidation state. This seems to be the reason why bismuth occasionally exhibits the oxidation state +1, in contrast to its lighter congeners in which the splitting of p orbitals into $p_{1/2}$ and $p_{3/2}$ is much smaller, Fig. 11.1. Bi^+ is not stable in aqueous solutions but, at least in the form of the salt $Bi(CF_3SO_3)$, is remarkably stable in dimethylthioformamide.

The elements from P to Bi in the trivalent state exhibit sp^3 hybridization with the lone pair at one of the corners of the tetrahedron. In pentahalogenides the hybridization is sp^3d or sp^2 with a 3c-4e axial bond (see Section 5.1) (though solid PCl_5 is actually ionic, consisting of tetrahedral $[PCl_4]^+$ and octahedral $[PCl_6]^-$). The low boiling temperatures of most of the tri- and penta-halogenides are evidence for high contributions from covalence to the bonding which, however, decreases down the Group. In oxides, oxoacids and oxoanions both P^{III} and P^V show sp^3 hybridization. The acids H_3PO_4, H_3PO_3, and H_3PO_2 are tribasic, dibasic, and mono-basic respectively – they are more informatively written $O=P(OH)_3$, $O=PH(OH)_2$, and $O=PH_2(OH)$. Phosphoric and phosphorous acids, and their respective anions, show tetrahedral coordination of oxygen around the phosphorus atom and contain, apart from four σ bonds, $p\pi-d\pi$ bonding arising from overlap of a d orbital on phosphorus with a p orbital on oxygen. In orthophosphoric acid (H_3PO_4 or $O=P(OH)_3$) the $p\pi-d\pi$ bond is localized on P=O, whereas in orthophosphate anions the $p\pi-d\pi$ bonding is delocalized over the hydrogen-free P⸺O groups. Orthophosphoric acid and its salts show a strong tendency to condense. By sharing two oxygen atoms PO_4 tetrahedra can either form chain polyphosphates with the general formula $[P_nO_{3n+1}]^{(n+2)-}$ (n = 1 to 16) or cyclic metaphosphates $[(PO_3)_n]^{n-}$ with $n = 3$ or more.

In contrast to nitrogen oxides, the oxides of P, As and Sb in both the +3 and +5 oxidation states form mixed clusters (see Section 6.2). For instance in the M_4O_6 oxides (M = P, As, Sb) the four M atoms are at the corners of a tetrahedron while the six oxygen atoms, each bonded to two M atoms, lie along the edges. The oxides of P, As and Sb have acidic properties while Bi_2O_3, because of large ionic radius of Bi^{III}, is basic. Only for bismuth in oxidation state +3 is there a true cationic chemistry, with some scanty evidence for the existence of Bi^{3+}aq in acidic aqueous solution. Proposed pK_a values of 1.6 to 2.0 for this putative aqua-cation suggest a strong acidity, analogous to that for Tl^{3+}aq, and are consistent with the strongly hydrolyzed and polymerized nature of aqueous Bi^{3+} solutions at most pHs. Numerous hydroxo-bridged polynuclear Bi^{3+} species have been proposed; the highly symmetrical $[Bi_6(OH)_{12}]^{6+}$ cation contains such an octahedral array of bismuth atoms.

Arsenic, antimony and bismuth in oxidation state +3 form complexes MX_4^- and MX_5^{2-} with halide ions, whereas in oxidation state +5 mainly MF_6^- complexes are formed (except for Bi). The structure of the MX_4^- complexes is a trigonal bipyramid with the lone electron pair in the equatorial position. The MX_4^- complexes show a strong tendency to form polynuclear species, for instance, two SbF_4^- anions form $Sb_2F_8^{2-}$ by sharing two fluorine atoms. Coordination of each of the Sb^{3+} cations then becomes tetrahedral bipyramidal with a lone pair in the axial position. In the

MX_5^{2-} complexes the lone pair of electrons occupies an axial position in the tetrahedral bipyramid, whereas the MF_6^- complexes are octahedral. Both types of complex are hypervalent. The PF_6^- anion is considered to be one of the least complexing anions. On the other hand SbF_5 is an extremely strong Lewis acid which in liquid HF enhances the acidity of the HF by forming $[H_2F][SbF_6]$.

Many P(III), As(III), and Sb(III) compounds are good donor ligands and also, through π-back bonding, capable of stabilizing low oxidation states. Such P and As donors, especially aryl and alkyl derivatives PR_3, $P(OR)_3$, and AsR_3, play an extremely important role in organometallic chemistry and in homogeneous catalysis.

11.3 COMPARISON WITH GROUP 5 ELEMENTS

The elements of Groups 5 and 15 differ tremendously in the structure of their respective solid phases. The Group 5 elements are cubic close-packed, whereas As, Sb and Bi form interacting double layers, with each atom having a coordination number of 3. The reason is the distinctly lower ionization potentials of Group 5 than of Group 15 elements. For instance, the sum of the first three ionization potentials equals 46.2 eV for Nb, but 52.5 eV for Sb. Elements of the two Groups also differ in their melting points, which are low for As, Sb and Bi, and high for V, Nb and Ta. Melting points decrease down Group 15 and increase down Group 5.

The Group 5 and 15 elements show the same +5 and +3 oxidation numbers. However, the stability of the +5 oxidation state increases down Group 5, whereas in Group 15 it is unstable for the heaviest member i.e. for Bi. The Group 5 elements also show the oxidation numbers +4 and +2, which are unknown in Group 15, except for occasional nitrogen compounds. The +4 oxidation state is particularly important for vanadium, in tetrahalides and in the form of the unusual but very stable VO^{2+} (vanadyl) ion with its extensive solid state and solution chemistry. Formally there is an analogy between VO^{2+} and NO^+ salts, but of course the oxidation states of the V and N differ. The +2 state is also important for vanadium, even though V(II) complexes, including $V^{2+}aq$, are strongly reducing. In organometallic derivatives vanadium shows oxidation numbers down to zero, -1, and as low as -3 in $[V(CO)_5]^{3-}$.

Probably the closest similarities between Group 5 and 15 elements are to be found in their oxidation state +5 oxoanions. Thus vanadium(V) resembles phosphorus(V) in forming ortho-, pyro- and meta-anions. Orthovanadates, like orthophosphates, contain discrete tetrahedral VO_4^{3-} ions, whereas pyrovanadates contain, again like pyrophosphates, dinuclear $[V_2O_7]^{4-}$ ions which consist of two VO_4 tetrahedra sharing a corner. Further condensation gives rise to a variety of oxoanions, including such species as $[V_3O_9]^{3-}$, $[V_4O_{12}]^{4-}$, and $[V_{10}O_{28}]^{6-}$. P^V (and other M^N) can be incorporated, as in the heteropolyanion $[PV_{12}O_{36}]^{7-}$. For V^V, as for P^V, MO_4 tetrahedra are never edge- or face-shared. Overall the oxoanion chemistries of phosphorus and vanadium show considerably more similarities than those of phosphorus and nitrogen. There are no vanadium analogues of the P–P-linked, as in diphosphate, $[O_3PPO_3]^{4-}$, (or S–S linked, as in dithionite or dithionate, Section 12.2) oxoanions. Neither are there, in lower oxidation states, vanadium analogues of the phosphite and hypophosphite anions with their P–H as well as P–OH and P=O bonds.

Trihalides are known for all Group 5 and 15 elements with all four halogens (except perhaps TaI_3), but rather fewer pentahalides exist. Pentahalides, even pentaiodides, are stable for Nb and Ta; for Bi only the pentafluoride is known. Such a distribution is consistent with the usual general pattern of stabilities of high oxidation states increasing down a d block Group, decreasing down a p block Group. For Group 5, in addition to the tri- and penta-halides almost all the conceivable tetrahalides are known, as are all the dihalides of vanadium. The elements of both Groups form a variety of halogenometallates $[MX_x]^{n-}$, though now the range and number of examples are considerably greater for the Group 15 elements (except nitrogen). Indeed for the Group 5 metals these are practically restricted to fluoroanions $[MF_x]^{n-}$, which are stable for V, Nb and Ta as they are for P, As and Sb. Whereas in Group 15 such species are the hexafluoroanions $[MF_6]^-$, in the case of Nb and Ta it is also possible to generate and isolate the penta- and octa-fluorometallates $[MF_7]^-$ and $[MF_8]^-$ under appropriate conditions.

There is relatively little simple hydrated cation chemistry in either Group 5 or Group 15. In Group 15 the only candidate is the elusive Bi^{3+}aq (see above, p. 112). In Group 5 there are V^{2+}aq and V^{3+}aq, as well as the VO^{2+}aq mentioned above, but as yet no simple mononuclear aqua-cations of Nb or Ta have been characterised. The nearest species so far reported are polynuclear mixed valence clusters such as $[Nb_3(\mu^3\text{-}Cl)(\mu\text{-}O)_3(H_2O)_9]^{4+}$, in which at least there are three water molecules bonded to each Nb.

12

Group 16

Table 12.1 – Fundamental properties

	O	S	Se	Te	Po
$R = \langle r_{np} \rangle$ / pm	65.5	110	123	144	155
I_1 / eV	13.6	10.4	9.7	9.0	8.4
A / eV	1.46	2.08	2.02	1.97	1.90
χ	3.4	2.6	2.5	2.1	1.9
bp (°C)	−183	718	685	990	962
mp (°C)	−218	113 [a]	217 [b]	450	254
$r_i(2-)$ / pm (CN 6)	140	184	198	221	230
$r_i(4+)$ / pm (CN 6)		37	50	97	94
$E°(0/2-)$ / V	+1.23	−0.45	−0.67	−1.14	−1.4

[a] For orthorhombic sulphur. [b] For grey selenium.

Oxygen, as the first element in the Group, differs from its heavier congeners, in accordance with the general rule. The differences in chemical properties result from its high ionization potentials, small radius and high electronegativity, which in turn are the effects of the singular properties of the $2p$ orbitals.

12.1 PROPERTIES OF OXYGEN

Because of the small radius of the oxygen atom, the O_2 molecule (dioxygen), with one σ and one $p\pi-p\pi$ bond, is strongly bonded ($\Delta H_{diss} = 498$ kJ mol^{-1}), although not so strongly as the N_2 molecule (O=O vs N≡N). The O_2 molecule has two unpaired electrons in the antibonding π orbitals, which are empty in dinitrogen (see Fig. 11.2) and is, therefore, paramagnetic. The dioxygen molecule forms complexes with metal ions. Of particular importance is the reversible reaction between the O_2 molecule and iron in haemoglobin and in myoglobin. Attachment of O_2 to the Fe^{2+} ion in deoxyhaemoglobin is accompanied by oxidation of divalent iron to Fe^{3+} and reduction of O_2 to the superoxide ion O_2^-. O−O bond distances in the sequence $O_2^{2-} \rightarrow O_2^- \rightarrow O_2 \rightarrow O_2^+$ (the first of importance in peroxides, e.g. Na_2O_2; the last obtainable in the form of e.g. its PtF_6^- salt) are 149, 128, 121, and 112 pm. These bond distances, v_{O-O} stretching frequencies, and bond dissociation energies of 204, 498, and 625 kJ mol^{-1} for O_2^{2-}, O_2, and O_2^+ all reflect the variation in the electron populations of the antibonding π orbitals (see Fig. 11.2). Bond orders are 1, 1½, 2, and 2½ in O_2^{2-}, O_2^-, O_2, and O_2^+ respectively. The oxygen atom combines with the hydrogen atom to form the OH radical which is a powerful oxidant. The HO_2 radical is also a strong oxidant; both radicals play an important role in initiating oxidative reactions in biological systems. Ozone, O_3, the second allotrope of oxygen, is a stronger oxidant than dioxygen and in this respect is second only to fluorine. The recently discovered third allotrope, O_4, is a fugitive gaseous species. However, infrared studies have suggested that solid oxygen at very high pressures may be built up from O_4 units, which confer a red coloration on this solid.

Oxygen is divalent and shows, as a rule, oxidation number –2 (–1 in Na_2O_2 and –½ in CsO_2). Only in OF_2 does oxygen show an oxidation number of +2, because fluorine is even more electronegative than oxygen. Owing to its high electronegativity oxygen forms the O^{2-} anion, when combined with electropositive elements. Despite the high bond dissociation energy dioxygen is very reactive and combines with almost all elements. Because of its high electronegativity oxygen, like fluorine, promotes formation of high formal oxidation states by metals e.g. Mn(VII) in MnO_4^-, Fe(VI) in FeO_4^{2-} and Os(VIII) in OsO_4. In connection with this ability one should take into account that the radius of the O^{2-} anion is only 140 pm (a small value for an anion) and that its charge is twice that of halide ions. The small radius of O^{2-} and its 2– charge result in high lattice energies of metallic oxides and high M–O bond energies in complexes. This makes up for the energy necessary to multiply ionize metal atoms. High element-to-oxygen bond energies are also characteristic of most of the large number of oxoanions, of p- and of d-block elements.

In the majority of oxygen compounds the hybridization of oxygen is sp^2 or sp^3 with two lone electron pairs in the vertices of a triangle or a tetrahedron, respectively. These lone pairs are readily shared with electron pair acceptors. Because of this property oxygen atoms in ethers, ketones, alcohols and, above all, in water molecules eagerly enter the inner coordination sphere of metal ions. The formation of four hydrogen-bonds by each water molecule is responsible for the unique structure of liquid water and ice.

Formation of $p\pi-p\pi$ bonds by the oxygen atom is greatly facilitated by its small radius ($R = \langle r_{2p} \rangle = 65.5$ pm). Such $p\pi-p\pi$ bonding is present in the >C=O group, where hybridization of the oxygen atom is sp^2. Because of the very small radius of the oxygen atom rotation about the O–O axis in the hydrogen peroxide molecule H_2O_2 is hindered by the electron pairs in the nearby O atoms. The hydrogen peroxide molecule is the smallest known molecule with hindered rotation. In contrast to S, Se and Te, oxygen catenates very poorly and forms only the relatively stable O_3 molecule, which contains two σ bonds and one delocalized π bond (a 3c-4e bond), and the very unstable O_4 molecule mentioned above.. The ease of formation of $p\pi-p\pi$ bonds by oxygen atoms hinders formation of longer chains. This is because in the O_3 molecule neither of the end oxygen atoms has an unpaired electron (they participate in the delocalized π bond), which is a necessary condition for chain propagation.

12.2 PROPERTIES OF S, Se, Te AND Po

Like the Group 15 elements the elements of Group 16 (except for polonium) do not form cations and their chemistry is mainly that of oxides, halides, oxo-acids and anions. This make changes within the Group more gradual and uniform, although the unexpectedly strong oxidising properties of selenic acid should be noted.

A very specific property of elementary sulphur is the formation of solid phases based on various catena-forms (see Section 6.1). Several tens of allotropic modifications of sulphur are known. Formation of polythionate $[O_3S-S_n-SO_3]^{2-}$ ($n = 0 - 4$)* and polysulphide S_n^{2-} ($n = 2 - 6$) anions provide other examples of the strong tendency of sulphur to catenate. Selenium and tellurium also catenate to form stable solids consisting of infinite chains of the respective atoms, and can co-catenate with sulphur in, for instance, $[O_3SSSeSSO_3]^{2-}$ and its tellurium analogue and in $[O_3SSeSeSO_3]^{2-}$.

In accordance with the general rule, the metallic character of the solid phase formed by the Group 16 elements increases down the Group. Sulphur is an excellent insulator, selenium a photoconductor, and tellurium a semiconductor, but polonium is a genuine metal. The energy gap between the highest filled band and the empty conduction band decreases from 1.84 eV for Se to 0 for Po. It is interesting that on going from Se to α-Po the ratio of the distances between the nearest and next-nearest neighbours increases from 0.67 to 1.0, and is accompanied by an increase of coordination number from 2 to 6. The increase in the ratio of distances up to 1.0 means that the bands broaden down the Group and finally overlap at polonium. The bands broaden because of increasing interaction between the chains.

Due to their high electron affinity the Group 16 elements show a tendency to complete their octet and thus form X^{2-} anions (oxidation state -2). Indeed, the Group 16 elements form sulphides, selenides, tellurides, and polonides with a wide range of elements, both metals and non-metals. It is interesting to see how the structures of 1:1 binary compounds vary with the position of the two constituents in the Periodic Table. As implied in Table 12.2, almost all 1:1 oxides have the rock salt structure, except when a cation such as Cu^{2+} (Jahn-Teller), Zn^{2+} (small; tetrahedral) or Hg^{2+} (linear coordination strongly favoured) imposes a different structure. But structural chemistry becomes more varied as one descends Group 16, with radii, covalency, and other factors leading to a greater range of structures. The 1:2 sulphides MS_2 have covalent structures of the pyrites (M = e.g. Fe, Co, Ni) or CdX_2 layer (M = e.g. Sn, Ti, Zr, Ta; see Chapter 13 for CdX_2 structures) type, but 2:1 selenides M_2Se (M = Li, Na, K) have the ionic antifluorite structure, with the relatively large Se^{2-} ions permitting eight M^+ cations to be nearest neighbours.

The tendency to form X^{2-} anions (oxidation state -2) decreases down the Group, which is shown by the variation in $E^\circ(0/2-)$ reduction potentials quoted in Table 12.1. The Group 16 elements from S to Po show, as a rule, even valencies from II

---

* The dithionate anion, $S_2O_6^{2-}$ (i.e. n = 0), is a useful precipitant for coordination complexes, tetrathionate, $S_4O_6^{2-}$ (i.e. n = 2), is the product of the familiar iodine-thiosulphate titration, and hexathionate, $S_6O_6^{2-}$ (i.e. n = 4), is the highest member of the series to have been structurally characterized.

Table 12.2 – Structures of selected 1:1 binary compounds of Group 16 elements.

	oxides	sulphides	selenides	tellurides	polonides
Mg	NaCl	NaCl	NaCl	ZnS (W)	NiAs
Ca, Sr, Ba	NaCl	NaCl	NaCl	NaCl	NaCl
Zn	ZnS (W)	ZnS (W; ZB)	ZnS (W; ZB)	ZnS (W; ZB)	
Cd	NaCl	ZnS (W; ZB)	ZnS (W)	ZnS (ZB)	ZnS
Hg	chain	chain	ZnS (ZB)	ZnS (ZB)	
Ti, V	NaCl	NiAs [a]			
Mn	NaCl	NaCl; ZnS (W)		NiAs	
Fe, Co, Ni	NaCl	NiAs	NiAs	NiAs	
Cu	~ PtS	contains CuS$_3$ and CuS$_4$ units			
Eu	NaCl	NaCl [b]	[c]		NaCl

[a] But ZrS has the NaCl structure.   [b] Also LaS, CeS, . . . HoS, and ThS, US, PuS.
[c] ScSe, YSe, and LuSe have the NaCl structure.

to VI and even oxidation numbers from −2 to +6. Exceptionally, sulphur shows odd formal oxidation numbers, +1 in S_2F_2 and +5 in S_2F_{10}, but according to the general rule the molecules are dimers with S–S bonds, see Section 5.4. Divalent sulphur shows sp^3 hybridization and forms the tetrahedral SCl_2 molecule where it displays an oxidation number of +2, and the tetrahedral H_2S molecule where it is in oxidation state −2. The remaining two corners in each tetrahedron are occupied by electron pairs. According to valence bond theory (VSEPRT[¶]) the elements from S to Po in the tetravalent state adopt sp^3d hybridization. For instance, SF_4 forms a (distorted) bipyramid with the lone pair in the equatorial position. When combined with oxygen the tetravalent sulphur atom is sp^2 (e.g. in SO_2) or sp^3 (e.g. in SO_3^{2-}) hybridized and forms $p\pi-d\pi$ bonds using its empty $3d$ orbitals. In contrast to the O_3 molecule the isoelectronic (with respect to valence electrons) SO_2 molecule has two π bonds, instead of one delocalized π bond. The two π bonds could be a $p\pi-p\pi$ and a $p\pi-d\pi$ bond. The oxygen atoms in the ozone molecule cannot form a $p\pi-d\pi$ bond, because their d orbitals are, in contrast to sulphur, unavailable.

In spite of its large radius the sulphur atom forms a $p\pi-p\pi$ bond with the oxygen atom relatively easily, because the radius of the latter atom is small. Formation of a $p\pi-d\pi$ bond between sulphur and oxygen is even easier because of favourable orientation of the d orbital lobes, which permits better overlap with p orbitals, Fig. 3.2. However, the bonding in SO_2 can also be explained without invoking

[¶] The full title for VSEPRT, *viz.* Sidgwick–Powell Gillespie–Nyholm Valence Shell Electron Pair Repulsion Theory, recognises the proposers, the chief propagandists, and the key feature of this aspect of valence bond theory which recognises the importance of non-bonding electron pairs on stereochemistry.

participation of sulphur $3d$ orbitals. For instance, one can assume that in the SO_2 molecule the $3p_z$ orbital of sulphur and the two $2p_z$ orbitals of oxygen combine to give three π molecular orbitals. Bonding would then be three-centre four-electron (3c-4e), i.e. the same as in the O_3 molecule. However, the very short S–O distance in SO_2 (143 pm) points to the formation of both $p\pi-p\pi$ and $p\pi-d\pi$ bonds.

In the hexavalent state and when combined with halogens, as in SF_6, SeF_6 and TeF_6 the hybridization is sp^3d^2 (octahedron). However, bonding in these hypervalent compounds (4 electrons beyond the octet) can be explained without employing nd orbitals (see Section 5.1). In compounds with oxygen, where sulphur shows the oxidation number +6, hybridization is sp^3 (sp^2 in SO_3) and $p\pi-d\pi$ bonds are formed. For instance the tetrahedral SO_4^{2-} anion has, apart from four σ bonds, two delocalized $p\pi-d\pi$ bonds formed by overlap of two d orbitals of sulphur with two p orbitals from two oxygen atoms. Instead, the triangular SO_3 molecule has, apart from three σ bonds, three (localized) $p\pi-d\pi$ bonds. As in the case of SO_2, the bonding in SO_3 can be explained using the concept of multi-centre bonding. To this end we can assume that four p_z orbitals on the four atoms in the SO_3 molecule combine to give four molecular orbitals: one bonding, two nonbonding, and one antibonding. The six electrons, which according to the π bonding model participate in three $p\pi-d\pi$ bonds, now populate the first three molecular orbitals.

Something of the range and variety of the oxoanion chemistry of sulphur is apparent in the preceding paragraphs. A further aspect is that of redox chemistry, since in peroxodisulphate, $S_2O_8^{2-}$, and in dithionite, $S_2O_4^{2-}$, this area provides a much used strong oxidant and powerful reductant respectively. But whereas dithionite, with its weak S–S bond (S atoms large), is a rapidly acting reductant, $S_2O_8^{2-}$ has a strong O–O bond (O atoms small) and often oxidises substrates very slowly. The variety of sulphur oxoanion chemistry also results in the interesting product difference consequent on 1-electron and 2-electron oxidations of sulphite, which give $S_2O_6^{2-}$ (dithionate) and SO_4^{2-} respectively.

As is the case with other p block Groups so also in Group 16 (from S to Po) the stability of the highest oxidation state decreases with increasing Z. However, Se(VI) is less stable than Te(VI), which together with instability of Po(VI) is an example of secondary periodicity. The relative instability of Se(VI) results from the presence of the filled $3d$ shell and that of Po(VI) from the presence of the filled $4f$ shell and the direct relativistic effect which stabilize the $6s$ electrons. Because p orbitals in heavy atoms are split, the configuration of the Po atom is $6s^26p_{1/2}^26p_{3/2}^2$. Relativistic stabilization of the $6p_{1/2}$ orbital (small radius and more negative energy) results in formation of the relatively stable oxidation state +2 for Po and formation of the Po^{2+} cation.

12.3 COMPARISON WITH GROUP 6 ELEMENTS

The Group 6 elements show rather little resemblance to the Group 16 elements. Chromium, molybdenum and tungsten, in contrast to the non-metallic Group 16 elements (except polonium), are metals with very high melting and boiling temperatures. What is common to the two Groups is the oxidation number +6, which is important for all the Group 16 elements (apart from oxygen) and for the three Group 6 metals. The Group 16 elements appear only in even oxidation states (except in a few dimers); the Group 6 elements show both even and odd oxidation numbers.

The two Groups of elements both form series of oxoanions. In this respect there is a particularly marked similarity between chromium and sulphur, which both form very stable MO_4^{2-} anions. Because of their almost identical sizes, the MO_4^{2-} anions of these two elements form sparingly soluble barium, strontium, and lead salts. Moreover both for Cr and for S two anions can share one oxygen, which results in the formation of dichromate, $Cr_2O_7^{2-}$, and disulphate (pyrosulphate), $S_2O_7^{2-}$. Chromate can undergo further condensation to give trichromate, $Cr_3O_{10}^{2-}$, then tetrachromate, $Cr_4O_{13}^{2-}$, in both cases by sharing corners of CrO_4^{2-} tetrahedra. Sulphur forms an analogous series, of which penta- and hexa-sulphates, $S_5O_{16}^{2-}$ and $S_6O_{19}^{2-}$, have also been characterized. In contrast to chromium, higher polynuclear sulphur oxoanions can also be formed by sulphur catenation, as in the polythionates $[O_3S-S_n-SO_3]^{2-}$ (see p. 117 above).

Chromates, and chromium trioxide, are strong oxidants, especially for organic materials, but molybdates and tungstates, like sulphates, are non-oxidizing. Oxides MO_2 exist for almost all the Group 6 and 16 elements; only for chromium is there an important sesquioxide, Cr_2O_3. The Group 6 metals and sulphur all have a profusion of oxides, but whereas most of the dozen or more well-characterized oxides of sulphur have the sulphur in low formal oxidation states the numerous non-integral oxidation state oxides of the Group 6 metals are intermediate in composition between MO_2 and MO_3. For molybdenum, for example, this group of intensely coloured mixed valence oxides includes Mo_4O_{11}, Mo_5O_{14}, Mo_8O_{23}, and higher nuclearity oxides.

The halides of Group 16 elements in all oxidation states are, as a rule, low melting and highly volatile compounds. The halides of the Group 6 elements in oxidation state +3 have more or less salt-like character, but MoF_6 and WF_6 have low melting and boiling temperatures and in this respect bear some resemblance to SF_6. Oxohalides of molybdenum(VI) and tungsten(VI) also resemble their Group 16 analogues, though the relative MO_2X_2/MOX_4 stabilities differ between the two Groups.

13

Group 17. The halogens

Table 13.1 – Fundamental properties

	F	Cl	Br	I	At
$R = \langle r_{np} \rangle$ / pm	57	97	112	133	145
I_1 / eV	17.4	13.0	11.8	10.4	9.6
A / eV	3.40	3.62	3.36	3.06	2.8
χ	4.0	3.16	2.96	2.66	2.2
bp (°C)	−188	−34	58.8	457	
mp (°C)	−220	−101	−7.2	114	
ΔH_{diss} / kJ mol^{-1} ^{a}	159	242	193	151	
$r_i(1-)$ / pm (CN 6)	133	181	196	220	~230
$E°(0/1-)$ / V	+2.87	+1.36	+1.09	+0.54	+0.2

^{a} For $X_2 \rightarrow 2\,X^-$

13.1 PROPERTIES OF FLUORINE

As is usual for p block elements, fluorine as the first element in the Group differs significantly from its heavier congeners. The specific chemical properties of fluorine result from the small radius of the atom, high ionization energy, very high electronegativity and small dissociation energy of the F_2 molecule (159 kJ mol^{-1}). In contrast to the N_2 and O_2 molecules the F_2 molecule is only singly bonded, because with two more electrons than in the dioxygen molecule the antibonding molecular π orbitals are filled (add two electrons to the molecular orbitals of dioxygen shown in Fig. 11.2). Apart from being singly bonded, the small dissociation energy of F_2 is probably also caused by strong repulsion between the nonbonding electrons in each of the two fluorine atoms. The repulsion is strong due to small size of the fluorine atom. Similar strong repulsion is also responsible for the low stability of the hydrazine H_2N-NH_2 and hydrogen peroxide $HO-OH$ molecules, which also contain small N and O atoms, respectively. Fluorine, due to its high electronegativity and the low dissociation energy of the F_2 molecule, is the most reactive element:

- It reacts even with noble gases, except for He and Ne.
- Many organic compounds ignite in an atmosphere of fluorine.

– It is a very strong oxidant, stabilizing high formal oxidation states of elements in such compounds as IF_7, ReF_7, PtF_6, TbF_4, BiF_5, AuF_5 and $KAgF_4$. Another example is the existence of stable PbF_4 and the instability of all other Pb^{IV} halides.

– Because of its very high electronegativity the fluorine atom, when bonded, efficiently withdraws electrons from other parts of a molecule. This is why, for example, trifluoromethanesulphonic acid, CF_3SO_3H, is a thousand times stronger than its analogue CH_3SO_3H.

However, it should be recognised that although fluorine is the most electronegative element, on any of the numerous scales of χ, it does not always bring out the highest oxidation state, sometimes ceding this to oxygen. Examples include manganese (MnO_4^- but only MnF_4), iron (FeO_4^{2-} but only FeF_3), and osmium (OsO_4 but as yet no definitive evidence for the long sought OsF_8). Also high oxidation state oxygen derivatives are sometimes more stable than their fluorine equivalents, as for the relatively stable ReO_4^- vs. the extremely fragile ReF_7 and periodates vs. IF_7.

Due to its very high ionization energy, whose value reflects the very small size of the atom, fluorine shows only the –1 oxidation number and never positive oxidation numbers. The small size of the F atom makes the C–F bond energy (486 kJ mol^{-1}) much higher than that of the C–Cl bond (332 kJ mol^{-1}). High bond energy together with unavailability of d orbitals in the fluorine atom is the main cause of the low reactivity of fluorocarbons. On the other hand the presence of d orbitals in the Cl atom facilitates hydrolysis of the C–Cl bond in many chlorocarbons. This is because a d orbital of the Cl atom can be used to form a transition state incorporating the incoming OH^- group. Similarly the remarkable unreactivity of NF_3 contrasts with the explosive tendencies and ready hydrolysis, to NH_3 plus HClO, of NCl_3.

Because of the small radius of the atom (and of the F^- ion) HF is a relatively weak acid, in contrast to the rest of haloacids. This is just one of numerous examples of the consequences of particularly strong hydrogen-bonding to fluoride (as to oxygen). Both the unavailability of d orbitals in fluorine and its high electronegativity prevent its forming oxoacids.

13.2 PROPERTIES OF Cl, Br, I and At

Chlorine and bromine at sufficiently low temperatures, and iodine at room temperature, form layered solid phases by stacking together the singly bonded X_2 molecules. Solid chlorine and bromine are electric insulators whereas iodine is a two-dimensional semiconductor, with an energy gap (E_g) of about 1.3 eV (for silicon E_g = 1.1 eV). Under a pressure of 170 kbar iodine becomes a genuine metallic conductor. One can presume that if astatine were not a radioactive element with a short half-life, $\tau_{1/2}$ = 8.1 h, it would be found to form a metallic phase. As explained in Section 6.5, the increase in metallic properties down the p block Groups or with pressure is the result of a decreasing energy gap between the filled bonding and nonbonding bands and the empty conduction (antibonding) band. It is interesting to note that on going from Cl_2 to I_2 the ratio of the distances between the nearest and next nearest neighbour atoms increases from 0.6 to 0.78. This means that the bonds broaden and show a tendency to overlap, which is a prerequisite for formation of a metallic phase.

Under normal conditions chlorine is a greenish gas, bromine a dark red liquid and iodine a black solid which sublimes to give violet vapour. The strong visible absorption spectra of Br_2 and I_2 arise from transitions from the highest-energy filled π-antibonding orbitals to the lowest-energy empty σ-antibonding molecular orbital. This suggests that the empty antibonding orbital may be low enough in energy to serve as an electron acceptor orbital, thus conferring on the Br_2 and, particularly, the I_2 molecules Lewis acid properties (see below). Elements of Group 17 can either involve their one unpaired electron in formation of a σ bond, as in the X_2 and interhalogen XY molecules, or to complete the octet and form the X^- anions. The formation of X^- anions is facilitated by high electron affinity and thus decreases down the Group, as shown by the standard reduction potential, $E^\circ(0/1-)$.

Besides the oxidation state of -1, chlorine, bromine and iodine in compounds with more electronegative fluorine and oxygen also display formal odd positive oxidation numbers $+1$, $+3$, $+5$ and $+7$. In the fluorides XF_3 (X = Cl, Br, I) and XF_5 (X = Cl, Br, I) the presumed hybridization of the central halogen atom is sp^3d (distorted trigonal bipyramid with two electron pairs in equatorial positions) or sp^3d^2 (distorted octahedron with one electron pair in one of the corners), respectively. The structures of these compounds are consistent with the VSEPR principle (cf. Section 12.2, p. 118). In the fluoride IF_7 hybridization should be sp^3d^3 and the coordination polyhedron the pentagonal bipyramid. All these binary interhalogen compounds are hypervalent, because there are 2, 4 or 6 extra electrons beyond the octet. As with other hypervalent molecules, bonding in the binary interhalogens can be explained without invoking the participation of nd orbitals (Section 5.1). To this end we can invoke the concept of the 3-centre 4-electron (3c-4e) bond, which removes the extra electrons from the valence shell of the central atom. Thus, for instance, in the XF_3 molecule there is one X–F 2c-2e bond and one 3c-4e F–X–F bond which holds the two extra electrons in the nonbonding orbital. An XF_5 molecule has one 2c-2e bond and two 3c-4e bonds, which latter hold the four extra electrons beyond the central atom's octet. Since in a 3c-4e bond two electrons link all three atoms (the other two electrons are nonbonding), bonding in hypervalent molecules is rather weak. This is the reason why all the interhalogen compounds are strong oxidants and halogenating agents.

In the oxoanions ClO_2^-, ClO_3^- and ClO_4^-, where chlorine displays oxidation numbers of $+3$, $+5$ and $+7$, respectively, hybridization is tetrahedral (sp^3) with two, one and no electron pair(s) at vertices. As in the case of sulphur oxoanions, the electronic structure of halogen oxoanions can be explained assuming that in addition to σ bonds $p\pi-d\pi$ bonds are formed by overlap of chlorine d orbitals with oxygen p orbitals. In the above-mentioned oxoanions of chlorine there are one, two and three delocalized $p\pi-d\pi$ bonds, respectively. The ClO_2 molecule, with one more electron than the SO_2 molecule, is paramagnetic. In the framework of three-centre bonding this extra electron occupies the antibonding orbital (cf. Fig. 5.1). Although ClO_2 is an odd-electron molecule it does not dimerize, probably because its extra electron is more efficiently delocalized than that in the NO_2 molecule. Bromine and iodine also form a number of stable oxoacids and corresponding oxoanions, the group of oxo-species of Cl, Br, and I forming a marked contrast with fluorine chemistry (cf. above).

Iodine and astatine, due to their small ionization energy, which as usual decreases down the Group, form complex cations with, e.g., pyridine, $[I(py)_2]^+$ ("positive iodine") and $[At(py)_2]^+$. Astatine probably also forms a simple hydrated At^+ cation. Because of the high electronegativity and high electron affinity of the halogen atoms and the relatively low dissociation energy of the X_2 molecules, not only fluorine but also the remaining halogens are chemically very reactive. Elementary chlorine, bromine and iodine, as well as chlorine and bromine in higher oxidation states, are strong oxidants. The unexpectedly strong oxidizing properties of perbromic acid, $HBrO_4$, difficulties in synthesizing perbromates, and the absence of the interhalogen compound BrF_7, in contrast to known IF_7, are examples of secondary periodicity. Just as in the case of arsenic the presence of a filled $3d$ shell stabilizes the $4s$ electrons in the bromine atom, which renders attainment of the highest possible oxidation state difficult. Stabilization of the $4s$ electrons in the bromine atom is demonstrated by the ε_{ns} orbital energies, which are -29.4, -27.8, -20.4 and -25.5 eV in the Cl, Br, I and At atoms, respectively (cf. Fig. 11.1).

A characteristic property of iodine is formation of brown-coloured polyhalide anions I_3^-, I_5^-, I_7^- and I_9^-. These are formed by interaction first of the I^- and then of the I_3^- anion with successive I_2 molecules:

$$I^- \xrightarrow{I_2} I_3^- \xrightarrow{I_2} I_5^- \xrightarrow{I_2} I_7^- \xrightarrow{I_2} I_9^-$$

These polyiodides are Lewis acid-base complexes in which I^- and I_3^- act as bases and I_2 as the acid. The I_3^- ion is symmetrical (I–I = 292 pm in Ph_4AsI_3) or nearly so (I–I = 284, 304 pm in CsI_3). Bonds in the higher anions usually differ in length, as detailed for I_5^- in Fig. 13.1. Bond distances may be compared with the I–I distance of 272 pm, and I···I distances to near neighbours of 350, 397, and 420 pm, in crystalline I_2. Bond distances in I_7^- are 274, 290, and 344 pm, which pattern indicates rather weak interactions between the two I_2 components and the I_3^- unit linking them. The most stable polyiodide is the I_3^- anion, whose formation is the reason why iodine is soluble in aqueous solutions of iodides. To complement these I_n^- (n odd) polyiodides there is another series I_n^{2-} (n even, = 4, 8, 16). Now two anions (I^- or I_3^-) are linked by I_2 molecules, as shown in Fig. 13.2. All these anions are hypervalent compounds and the excess of electrons is removed by formation of electron-rich 3c-4e bonds.

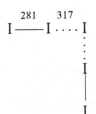

Fig. 13.1 Structure and bond distances (pm) for the I_5^- anion.

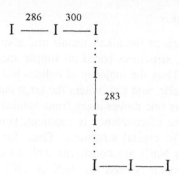

Fig. 13.2 Structure and bond distances (pm) for the I_8^{2-} anion.

The halogens also form polycations such as I_2^+, I_3^+, I_4^+, and I_5^+. The I_2^+ cation is paramagnetic and intensely coloured. Its I–I bond distance is 256 pm, significantly shorter than I–I = 272 pm in the I_2 molecule. The bond distances for I_5^+ are shown in Fig. 13.3. The I_2 molecule, which is an electron acceptor (Lewis acid), forms adducts not only with the I^- anion but also with other electron donors, which shifts the absorption band of the I_2 molecule toward higher frequencies. This change in absorption is responsible for the brown colour of iodine in electron-donor solvents such as alcohols, ethers and aromatic solvents, in contrast to the violet colour in solvents which are not electron donors, as for example aliphatic hydrocarbons or CCl_4.

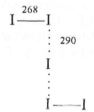

Fig. 13.3 Structure and bond distances (pm) for the I_5^+ cation.

The halides also form mixed polyhalide anions, such as BrF_4^-, ICl_2^-, and even $BrCll^-$, which are also hypervalent species. The structures of these anions conform to VSEPRT, with, for example, ICl_2^- containing linear Cl–I–Cl, with the three lone pairs of electrons residing at the triangular vertices of a trigonal bipyramid, and ICl_4^- having a square-planar array of chlorides and two lone pairs at the vertices of the six-electron-pair octahedron. In ICl_4^- there are, formally, two 3c-4e bonds which hold the four extra electrons beyond the octet of the iodine atom. Salts of these anions can be made readily from appropriate halides and interhalogen compounds, e.g.

$$CsCl + ICl_3 \rightarrow Cs[ICl_4] \ .$$

Preparation is easier, and the product more stable, for large cations, such as Cs^+ or R_4N^+. This is because there is less loss of lattice energy on going from halide to polyhalide for a large cation. Lattice energy also controls the product of decomposition, which is the salt of higher (highest) lattice energy. Thus on warming $Cs[ICl_2]$ gives $CsCl + ICl$ rather than $CsI + Cl_2$; similarly $KBrF_4$ gives $KF + BrF_3$.

13.3 STRUCTURES OF IONIC HALIDES

The halides of electropositive elements such as the alkali metals and alkaline earth metals tend to have, as one would expect, structures based on simple space-filling, radius ratio, and electrostatic principles. Thus the majority of halides MX have the 6:6 rock salt structure, except for CsCl, CsBr, and CsI, where the large caesium ion favours the 8:8 CsCl structure. However as one moves away from halides of Group 1 and 2 metal cations an increasing degree of covalency is apparent, being shown first in thermochemical properties, then in crystal structures. Thus, for example, Born-Haber and related cycles applied to NaCl are consistent with an essentially fully ionic structure, but if such an approach is used for AgX or TlCl it can be estimated that the basic ionic interactions are supplemented by significant covalent contributions to bonding. Estimates of these contributions for silver(I) halides increase from about 30 kJ mol^{-1} for AgF to about 120 kJ mol^{-1} for AgI. As the ionic model used is self-compensating, the covalent contributions may well be considerably larger.

Whereas halides such as AgX and TlCl retain the essentially ionic lattice of NaCl, other chlorides, bromides, and iodides have structures which clearly demonstrate the effects of covalency. Thus $CdCl_2$ and CdI_2 have structures which can be described either in terms of close-packed halide ions with alternate layers of octahedral holes vacant and occupied by Cd^{2+} ions – a distinctly non-electrostatic arrangement as it contains adjacent sheets of halide anions – or as layers of CdX_2 molecules.

Comparison of the structures of NH_4F and of the other ammonium halides reveals the importance of hydrogen-bonding to fluoride, for NH_4F has an open wurtzite (4:4) structure, the others the 8:8 CsCl structure. Born-Haber cycle calculations on NH_4F permit an estimate of about 15 kJ mol^{-1} to be made for each H····F····H bond in this salt.

14

Group 18 (0). The noble gases

Table 14.1 – Fundamental properties

	He	Ne	Ar	Kr	Xe	Rn
$R = \langle r_{np} \rangle$ / pm	49	51	88	104	124	137
I_1 / eV	24.6	21.6	15.8	14.0	12.1	10.7
Promotion energy / eV [a]	19.8	16.6	11.5	9.90	8.30	6.80
bp (°C)	−269	−246	−186	−152	−107	−62
Solubility in water [b] / cm^3 kg^{-1}	8.7	10.1	31.4	56.1	98.2	208

[a] $1s^2 \rightarrow 1s^1 2s^1$ for He, $np^6 \rightarrow np^5 (n+1)s^1$ for the rest. [b] At 25 °C; 0.1 MPa.

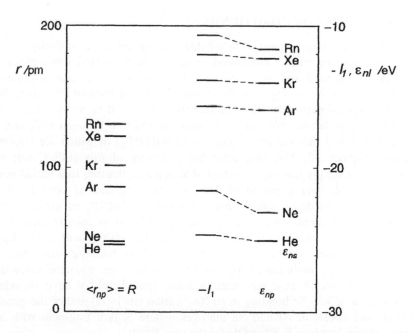

Fig. 14.1 Radii of atoms, $R = \langle r_{np} \rangle$, energy of np orbitals and first ionization potentials of noble gases.

14.1 INTRODUCTION

The characteristic feature of the noble gases is that their ionization energies and np^6 $\rightarrow np^5(n+1)s^1$ or $np^6 \rightarrow np^5nd^1$ promotion energies are very high. However, it should be noted that ionization potentials and promotion energies decrease very rapidly down the Group. As in other p-block Groups changes in basic properties of atoms with Z, Fig. 14.1, are not uniform. One can see in Table 14.1 and Fig. 14.1 close similarity between He and Ne and a sudden change in all fundamental properties between Ne and Ar. The similarity between He and Ne with respect to the radius of the atom and the energy of its outermost orbital is a consequence of high effective nuclear charge ($Z_{eff} = 5.85$) acting on the $2p$ electron in the neon atom. High Z_{eff} is, in turn, the result of incomplete shielding of the nuclear charge by the rest of the p electrons. The similarity between helium, which is usually thought to be the first element in Group 18, and neon, which is the next element, is contrary to what we know from previous Chapters about the unique properties of each first element in Groups 13 to 17. However, this discrepancy disappears when helium, in accordance with its $1s^2$ configuration, is placed in Group 2 as its first member. With neon as the first element in the Group 18 the gap between neon and argon confirms then the general rule concerning changes between the first and the next element in Groups of p block elements. The reason for the gap is repulsion between the $3p$ and $2p$ subshells in the argon atom (see Section 3.2). Table 14.1 shows that non-uniform changes in the radii of atoms are paralleled in the sequence He$\rightarrow$Xe by similar non-uniform changes in boiling point and solubility in water.

14.2 FORMATION OF COMPOUNDS

For a long time it was a common belief that noble gases are chemically inert. This belief was disproved when a compound thought to be $Xe[Pt^VF_6]$ was synthesized in 1962. This compound and its dioxygenyl analogue $O_2[Pt^VF_6]$ can be obtained by reaction of either Xe or O_2 with the extremely strong oxidant PtF_6 (see Section 15.6). One should notice that the first ionization potential of Xe is almost equal to that of the O_2 molecule. We now know that the reaction between PtF_6 and Xe is more complex and that the end product is $[XeF][Pt^VF_6]$ in which Xe displays the oxidation number +2. Not long after the discovery of this $[PtF_6]^-$ salt it was discovered that simply leaving a mixture of xenon and fluorine in a sealed vessel in sunlight resulted, over a period of days, in the formation of crystals of XeF_2. Subsequently it was demonstrated that irradiation of $Cr(CO)_6$ in matrices of solid argon, krypton or xenon gave unstable products of different colours, indicating the formation of species $ArCr(CO)_5$, $XeCr(CO)_5$, and $KrCr(CO)_5$ with some degree of noble gas – chromium interaction, i.e. bonding. This bonding may arise from polarizability of the noble gas atom, which may in this way donate electron density to the metal centre. Recent advances in noble gas chemistry have included the demonstration of Xe–Xe bonding in an Xe_2^+ cation (in 1997) and of the generation of HArX and HXeX in Ar and Xe matrices, where X is a fragment with a high electron affinity, such as F, Cl or OH (in the year 2000).

Due to high ionization and promotion energies noble gases enter into more or less stable chemical combinations only with most electronegative elements such as fluorine and oxygen. Because the Group number is even, noble gases show even formal oxidation numbers, ranging from +2 to +8. With decreasing ionization potentials down the Group the ability to react chemically increases. The threshold of significant chemical activity is reached at Kr. The chemical activity of Xe is much greater and that of Rn should be greater still, but has been only occasionally studied. We know, for instance, that radon reacts with fluorine and with halogen fluorides to form RnF_2. In conducting solvents, such as BrF_3, this is believed to dissociate, giving RnF^+, perhaps even Rn^{2+}. There is also evidence for formation of cationic species such as RnF^+ in an SbF_6^- salt. Although it has become customary to view compounds of this type as ionic, often emphasising this by writing their formulae in the form $[XF]^+[MF_6]^-$, there are in most cases large covalent contributions to bonding between the noble gas "cation" and fluoro-metallate "anion". Thus the XeF_2 adduct with SbF_5 contains a very distorted SbF_6 "octahedron", and can almost equally well be written as $XeF_2.SbF_5$, with a shared fluoride, *viz.* $F–Xe\cdots F\cdots SbF_5$. However, ionic formulations are in order for such adducts as $2XeF_6.PdF_4$, $2XeF_6.NiF_4$, and $XeF_6.AgF_3$, since these contain XeF_5^+ units and essentially octahedral PdF_6^{2-} and NiF_6^{2-}, and square-planar AgF_4^-, anions respectively. Complementarily to this F^- donor behaviour, xenon hexafluoride can act as an F^- acceptor, forming compounds of the XeF_7^- and XeF_8^{2-} anions on reaction with alkali metal fluorides (except Li) or NOF. As a further variant on donor properties in this group of compounds, we mention the suggestion that in $XeF_6.SbF_5$ the two moieties are bonded by donation of the electron pair on xenon to the SbF_5.

Xenon forms three fluorides: linear XeF_2, square-planar XeF_4 and distorted octahedral XeF_6. These compounds are hypervalent and incorporate 3-centre 4-electron bonds. In XeF_2 with its one 3c-4e bond the two xenon electrons and two fluorine electrons fill the bonding and nonbonding orbitals. In XeF_4 we have two, and in XeF_6 three, 3c-4e bonds. In this way the two, four or six electrons beyond the octet are removed from the xenon atom. Formation of XeF_2 and XeF_4 can also be discussed in the framework of valence bond theory. One can assume that one or two electrons in the Xe atom are promoted from the $5p$ to the $5d$ subshell and that the s, p and d orbitals hybridize to give either a trigonal bipyramid (sp^3d hybridization) or a square bipyramid (sp^3d^2 hybridization). In the trigonal bipyramid of the XeF_2 molecule fluorine atoms occupy axial positions and the three lone pairs equatorial positions. In the XeF_4 molecule the two lone pairs take axial positions in the square bipyramid. These XeF_2 and XeF_4 structures are analogous to those of ICl_2^- and ICl_4^- (see preceding Chapter), and are compatible with VSEPRT. The XeF_6 molecule perhaps oscillates between an octahedral structure (three $3c-4e$) bonds and a pentagonal bipyramid with the lone electron pair at one of the vertices. Possible bonding properties of this lone pair were mentioned at the end of the previous paragraph, in relation to the adduct $XeF_6.SbF_5$.

Xenon fluorides, like their interhalogen analogues, are powerful fluorinating agents and strong oxidants, behaviour which can be attributed to their 3c-4e bonds or participation of d orbitals (which requires promotion energy). Thus, for instance,

XeF_2 oxidizes water to oxygen. XeF_4 and XeF_6 are violently hydrolyzed by water, giving the dangerously explosive XeO_3 (of pyramidal structure). The octahedral XeO_6^{4-} anion is known; its barium salt is relatively stable. In XeO_6^{4-} and in the unstable XeO_4 molecule (tetrahedral, since there is no lone electron pair on the Xe atom) xenon displays a formal oxidation number of +8.

Ar, Kr and Xe readily form clathrates. These are substances formed by combining two stable compounds, or a stable compound with an atomic element, without any chemical bond between the two components but with only weak van der Waals interactions. Formation of clathrates may occur when one of the compounds, the so-called host, crystallizes in an open structure containing holes, cavities or channels in which the molecules of the second component, the guest, can be trapped. A condition for clathrate formation is matching the empty spaces in the host to the guest molecules or atoms. The guest cannot be too small, because then it can escape between the atoms of the host lattice. On the other hand the size of the guest molecule or atom cannot be too large, otherwise the host lattice would be strongly deformed and the system would become thermodynamically unstable. Clathrates are non-stoichiometric compounds in which the guest to host ratio depends on the conditions of preparation. An example of clathrate formation by Ar, Kr and Xe are compounds with β-quinol as the host. Formation of these clathrates requires pressures above 10 atm. The three noble gases also form clathrates with water at pressures greater than one atmosphere. Clathrates of this kind are usually called gas hydrates. In the structure adopted by noble gas hydrates the unit cell contains 46 molecules of water connected by hydrogen-bonds to form six larger and two smaller cages. Filling of cages depends on pressure and temperature and is usually incomplete. It should be noted that helium and neon do not form clathrates because of their small size. Interstitial noble gas – fullerene compounds, of which the first fully characterized example was ArC_{60}, form a link between clathrates and the noble gas compounds described in the preceding paragraphs.

15

Transition elements

15.1 GENERAL CHARACTERISTICS

Elements which have partly filled $3d$, $4d$ or $5d$ shells are called transition elements, whereas those with a partly filled $6d$ shell are commonly called transactinide elements.* According to this definition the Group 11 and 12 elements are not transition elements, because they have the d^{10} configuration. However, we can adopt a broader definition and also include as transition elements those that have a partly filled d shell in any of their commonly occurring oxidation states. According to the broader definition coinage metals are transition elements, since Cu(II) and Ag(II) have the d^9 configuration and Au(III) the d^8 configuration. The Group 12 elements are still not transition elements although chemically they frequently resemble transition elements. It should be noted that many authors also call lanthanides and actinides transition elements, a practice that we do not follow in this book.

All transition elements are metals with electronegativities in the range from 1.3 to about 2.2. Hence, some of them dissolve in non-oxidizing acids, while others require oxidizing acids or mixtures of both as e.g. aqua regia. Electronegativity and ionization potentials increase in each row from left to right, conferring on the end-elements in each series particularly noble character. On the other hand early members, if heated, react directly with most non-metals and in the case of titanium even with nitrogen. Some typical products, structures, and properties are mentioned in Section 15.4 below. In spite of their high reactivity toward oxygen the Group 4 and 5 elements show outstanding resistance to corrosion, which is due to formation of a dense, adherent oxide film.

Transition elements show a strong tendency to form molecules with metal-to-metal bonds as in e.g. $(OC)_5Mn–Mn(CO)_5$ or in clusters. A unique property of transition elements is formation of multiple bonds, in particular the quadruple bond, between metal atoms, which involves d orbital participation. The quadruple bond is a combination of one σ, two π and one δ bond. The σ bond, which is the strongest, results from overlap of the d_{z^2} orbitals on two metal atoms, the two π bonds, which are next in effectiveness, from interaction between the two pairs of d_{yz} and d_{zx} orbitals. Either the $d_{x^2-y^2}$ or the d_{xy} orbital from each metal atom combine to give the δ bond which, because of the small overlap, is the weakest. An example of such quadruply bonded metal atoms is the dimeric $[Cl_4Re\equiv ReCl_4]^{2-}$ ion.

---

* In this Chapter the general notation for d orbitals is nd, hence the higher-lying s orbitals are denoted $(n+1)s$ and the lower-lying p orbitals np.

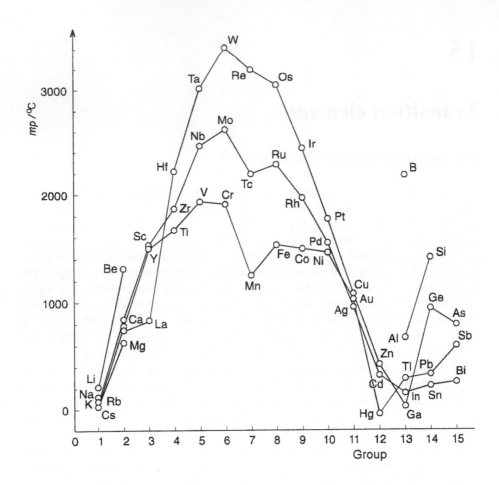

Fig. 15.1 Melting points of *s*, *p* and *d* block metals, and of B, Si, Ge, As and Sb.

15.2 THE METALLIC PHASE

Transition elements form metallic phases, which generally show very high atomization enthalpies, Fig. 6.3, and melting points, Fig. 15.1. In spite of the high atomization energies of transition elements the energy for a single M–M bond in the metallic phase is not exceptionally high. In the cubic and hexagonal closest packing which very frequently occurs in transition metals each atom has 12 nearest neighbours, hence even for osmium (CN 12) with its high atomization enthalpy of 790 kJ mol^{-1} the energy of the Os–Os bond is only about 132 kJ mol^{-1}. This should be compared with the atomization enthalpy of diamond (CN 4), 711 kJ mol^{-1}, and the C–C bond energy in diamond, 356 kJ mol^{-1}. The reason for the low energy of metallic bonding is its electron-deficient character. Thus, the high atomization enthalpies, and melting and boiling points of most transition metals are not the effect of bond strength but result rather from high coordination numbers (8 or 12) which, in turn, can be attained due to participation of d orbitals in bonding.

Transition elements show much higher enthalpies of vaporization, melting points and boiling points than s and p block metals. For instance, the atomization enthalpies of tungsten and potassium, which both crystallize in a cubic body-centred lattice (CN 8), are 849 and 89 kJ mol^{-1}, respectively. The main reason is the large number of bonding electrons in most transition elements in comparison with s block metals. Simple calculations show that the number of electrons per bond is 3/2 in tungsten and only 1/4 in potassium. Moreover, because of the smaller metallic radius the density of mobile electrons is much higher for tungsten than for potassium, which makes the ions more rigidly bound in the tungsten lattice. Instead, the difference between transition elements and p block metals with respect to atomization energy originates in higher ionization potentials for the latter, see Section 9.4.

The band structure of transition metals is more complex than that of s and p block metals and consists of a broad s/p band and a narrow d band. With respect to energy the d band is located approximately in the middle of the s/p band. Moreover, depending on lattice symmetry some d orbitals mix with s and p orbitals along certain crystallographic directions and confer some d character on the s/p band. The energy spread in the d band is narrow, because d orbitals do not extend to large distances from the atom – they are well within the s orbitals – which restricts their interactions with neighbouring atoms. As the transition series is traversed the d band narrows systematically. i.e. the d electrons become more and more localized as the result of contraction of d orbitals with increasing effective nuclear charge.

The melting points of the transition elements, like their atomization enthalpies, initially increase with increasing number of d electrons and then decrease. This is because up to the half-filled $s+d$ band, i.e. up to six electrons per atom (which happens on reaching Group 6), only bonding orbitals are filled, whereas subsequent electrons fill antibonding orbitals. However, in the first and second transition series the atomization enthalpy decreases between the total electron count from five to six, and decreases unexpectedly strongly between Cr and Mn, Fig. 6.3. This effect probably results from inefficient mutual screening of d electrons from increasing nuclear attraction. Inefficient screening makes ionization potentials related to the d subshell (I_3 and higher ionization potentials) increase markedly from d^1 to d^5 and from d^6 to d^{10} configurations, see Fig. 15.2. Because of increasing ionization potentials the atomization enthalpy decreases (see Section 9.4) on approaching the middle of a transition series, in spite of the energy gained from filling only bonding orbitals in the $s+d$ band. One can also say that because of decreasing orbital energy, i.e. increasing ionization potentials, the d orbitals tend to localize, which decreases their overlap on adjacent atoms and makes metallic bonding weaker. The dependence on ionization potentials also explains the increase of the atomization energy between Mn and Fe, parallel to decreasing I_3. The reason why the irregularity in the atomization enthalpy is absent from the third row and is much less pronounced in the second row may be the much larger radial extent of $5d$ and $4d$ than of $3d$ orbitals.

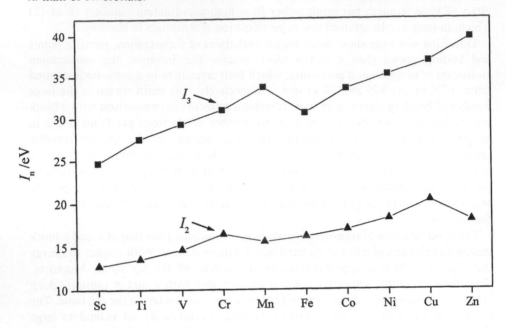

Fig. 15.2 Ionization potentials I_2 and I_3 for $3d$ elements.

15.3 OXIDATION STATES

For transition elements with the $d^n s^2$ configuration, i.e. for most elements in the first and third series, the +2 oxidation state as the lowest possible state (apart from some organometallic compounds) seems to be a natural choice. This is because detaching nothing but two s electrons exposes the very low lying np or the low lying e_g subshell for metal–ligand interaction in octahedral low-spin and high-spin complexes, respectively (see Section 15.5). The very strong or strong metal ion–ligand interaction, which is the result of higher charge and more or less short ligand-to-metal distance, prevails over the expenditure of energy necessary to remove the second s electron. In this respect transition elements with the s^2 configuration resemble Group 2 elements (see Section 5.2).

The elements of the second transition series, in spite of having mainly the $d^{n+1}s^1$ configuration, also exhibit the lowest possible oxidation number of +2, except for Ag whose usual oxidation number is +1. The reason is that after detaching the first electron from the $5s$ shell, removal of the next electron from the $4d$ shell requires an energy only slightly greater than that necessary to detach the second s electron from atoms of the first transition series. Two factors are responsible for the similarity of their second ionization potentials, I_2, in spite of electrons being detached from different subshells. The first factor is that repulsion between $3d$ and $4d$ subshells destabilizes the $4d$ subshell, which facilitates detachment of the $4d$ electron (see

Section 15.6). The second factor, acting in the same direction, is that one electron more in the d subshell (configuration $4d^{n+1}5s^1$ versus $4d^45s^2$) increases repulsion between d electrons which makes their orbital energy less negative.

In spite of their s^2 configuration the lowest stable oxidation state of Group 3 and 4 elements is not +2 but +3 and +4, respectively. The reason is the relatively low third ionization potentials, I_3, for Group 3 and relatively low third and fourth ionization potentials, I_3 and I_4, for Group 4 elements, Table 15.1. Because of these low I_3 and I_4 values removal of one or two d electrons, in addition to two s electrons, from Group 3 or Group 4 element atoms, respectively, does not require much energy. This energy is compensated by strong interaction of the highly charged M^{3+} or M^{4+} cations with counter-ions in a lattice, with water molecules, or with ligands.

Chemical experience shows that the stability of the +2 oxidation state increases as a rule in the $3d$ series from vanadium to copper. A similar trend can be also found for $4d$ and $5d$ elements which, in general, are reluctant to show the +2 oxidation state in classical complexes (see Section 15.6). The reason for the increasing stability of the lower oxidation state across the $3d$ transition series is that the ε_{4s} orbital energy and the related ionization potentials I_1 and I_2 change across the series more slowly than the ε_{3d} orbital energy and the related ionization potentials I_3 and I_4 (see Table 15.1, Fig. 15.2 and Fig. 2.3). The orbital energy ε_{4s} decreases more slowly than ε_{3d} because $4s$ electrons are shielded from nuclear attraction by $3d$ electrons which populate a subshell of much smaller radius. The increasing energy difference between $(n+1)s$ and nd orbitals and the resulting gap between the second and third ionization potentials, I_2 and I_3 (see Fig. 15.2), makes the +2 oxidation state more stable at the end of each series.

Table 15.1 – Changes in orbital energies and ionization potentials across the $3d$ series

	Sc	Ti	Mn	Ni
ε_{4s} / eV	-5.7	-6.1	-6.9	-7.7
$I_1 + I_2$ / eV	19.4	20.4	23.1	25.8
ε_{3d} / eV	-9.1	-10.8	-15.0	-18.5
I_3 / eV	24.8	27.5	33.7	35.3
I_4 / eV	(73.5)	43.2	51.2	54.9

After attaining the +2 oxidation state, the oxidation number of transition elements, in contrast to p block elements, frequently changes by one unit only. This is because removal of more than one electron from the d subshell does not expose a deeper-lying shell for metal-ligand contact. One should notice that in most octahedral low-spin (LS) complexes the radius of the central metal ion appears to depend on the radius of the outermost p subshell. The reason is that the t_{2g} orbitals project between the ligands, whereas the e_g orbitals which face the ligands are empty when the number of d electrons is < 6. Figs. 15.3 and 15.4 show that the

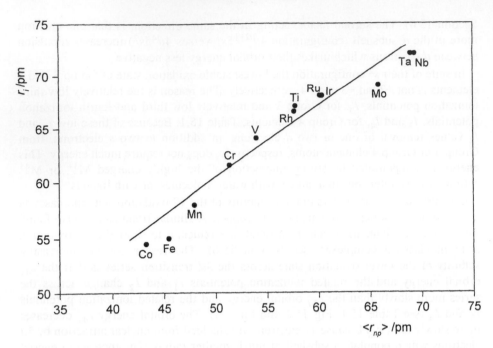

Fig. 15.3 The dependence of the M^{3+} ionic radius of d elements, r_i (LS; CN 6), on the radius of the np subshell, $\langle r_{np} \rangle$.

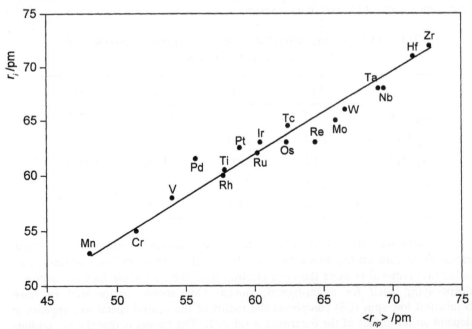

Fig. 15.4 The dependence of the M^{4+} ionic radius of d elements, r_i (LS; CN 6), on the radius of the np subshell, $\langle r_{np} \rangle$.

radii of M^{3+} and M^{4+} ions in low-spin complexes are a good linear function of the np shell radii (compare with the data for s and p block ions in Fig. 3.6). That the radii of M^{3+} and M^{4+} ions in low-spin complexes are indeed determined by the radius of the np subshell is shown by the small difference between r_i and $\langle r_{np} \rangle$ (again compare Fig. 3.6). With the metal–ligand distance not decreasing appreciably with removal of a d electron (compare plots in Figs. 15.3 and 15.4) the energy of the metal–ligand bond increases mainly due to increasing cation charge. On the other hand the increase of oxidation number requires additional ionization energy. The two factors more or less counterbalance each other, so that two oxidation states differing by one d electron do not differ very much in stability. This would not be the case if the ionic radius were decreasing abruptly with increasing oxidation number due to exposing a deeper lying subshell (see Section 5.2). Thus, the constant radius of the outermost shell in the ion, irrespective of the oxidation number higher than +2, is the main reason why transition elements are rich in oxidation states and why their oxidation numbers can change by one unit. The highest formal oxidation number increases in each transition series from 3 at the beginning of the series to 8 in the middle, e.g. in osmium tetroxide, and then decreases. The change in the oxidation number across each transition series is a compromise between increasing number of valence electrons and increasing ionization potentials, Fig. 15.2 and Table 15.1. The increase in ionization potentials across each series is due to the increase of the effective nuclear charge acting on the valence electrons.

The marked change in ionization potentials going across d-block rows has its reflection in aqueous solution chemistry (Fig. 15.5), though the relation between ionization potentials and redox potentials is not simple, as such parameters as hydration energies are also important. Sc^{2+}aq and Ti^{2+}aq are too unstable to have more than a transient existence, V^{2+}aq and Cr^{2+}aq are strong reductants, Fe^{2+}aq is a weak reductant, and Ni^{2+}aq has no reducing properties. The drop in $E^{\circ}(3+/2+)$ between Mn and Fe reflects the drop in I_3 (see Fig. 15.2) which, in turn, results from the difference in exchange energies (Section 15.6).

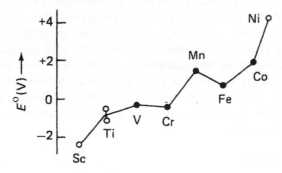

Fig. 15.5 The variation of M^{3+}/M^{2+} reduction potentials across the first row of the transition metals; ● represents an accurate value, ○ an approximate estimate.

15.4 BINARY COMPOUNDS, SALTS, HYDRATES, and AQUA-IONS

15.4.1 Binary compounds

The transition elements form, generally by direct combination and often only at high temperatures, a range of binary comounds with many of the non-metallic elements. There are a great range and variety of hydrides, borides, carbides, nitrides, oxides, second row equivalents such as phosphides and sulphides, and halides. The natures of these binary compounds range from ionic to covalent; binary oxides and fluorides tend to be ionic, other halides and sulphides tend to be significantly more covalent in character. The structures of the predominantly ionic compounds are generally consistent with radius ratio rules. The first-row oxides MO have the $6:6$ rock salt structure, the fluorides MF_2 have the $6:3$ rutile structure, contrasting with the $8:4$ fluorite structures of the fluorides of the larger Ca^{2+}, Sr^{2+}, and Ba^{2+} cations. The mixed valence oxides Fe_3O_4 and Co_3O_4 exhibit an interesting conflict between size (ionic radii) and electronic (Crystal Field) effects. Both of these M_3O_4 oxides have the spinel type of structure based on cubic close packed oxide ions, but whereas the cobalt compound has Co^{2+} in the tetrahedral holes and Co^{3+} in the octahedral holes, the iron compound has Fe^{3+} in the tetrahedral holes while the octahedral holes contain equal numbers of Fe^{2+} and Fe^{3+} ions. The key determining feature is the high Crystal Field stabilization of the d^6 Co^{3+} cation in the octahedral environment.

There is also a group of binary compounds, including many hydrides, borides, carbides and nitrides, classified as interstitial or metallic compounds. The term 'interstitial' was originally used as it was believed that the small atoms of the non-metallic element simply entered interstitial sites in the metal lattice, but it subsequently became apparent that inserting the non-metal atoms often affected the structure of the host metal. Moreover, despite the implication of weak bonding in the term 'interstitial', many of these compounds, e.g. tungsten carbide, are exceedingly hard. This hardness suggests very strong metal–non-metal interaction, as do their often small enthalpies of formation, which must represent a balance between the large enthalpy of dissociation of, e.g., dinitrogen or of atomization of solid carbon and enthalpically favourable bonding within the binary compound. In contrast to the ionic binary compounds, the structures of interstitial (metallic) binaries tend to have characteristics reminiscent of solids with significant covalent contributions to bonding. Thus such compounds as WC, WN, RuC and MoP have structures of the NiAs type, with $6:6$ trigonal-prismatic/octahedral coordination rather than the ionic all-octahedral $6:6$ NaCl structure. Similarly Co_2N has a $CdCl_2$-like layer structure rather than the rutile or fluorite structures characteristic of predominantly ionic solids.

15.4.2 Salts and hydrates

The d-block elements, especially those of the first row, form a large number of salts, with a great variety of oxoanions, halogenoanions, and polyatomic anions such as thiocyanate and azide, as well as the halides already mentioned in the preceding sub-section. Many of these salts crystallize from aqueous media as hydrates; in a number of cases different hydrates may be obtained according to conditions. Often the structures of these hydrates consist of hydrated cations $[M(H_2O)_6]^{n+}$ and anions, as for example in $[Cr(H_2O)_6](ClO_4)_3$. The anions may also be hydrated, for example by hydrogen-bonding between water and oxygen of an oxoanion – nickel sulphate heptahydrate is composed of $[Ni(H_2O)_6]^{2+}$ cations with the seventh water hydrogen-bonded to sulphate. Hexahydrates do not always contain $[M(H_2O)_6]^{n+}$ cations – $FeCl_3.6H_2O$ is $[Fe(H_2O)_4Cl_2]Cl.2H_2O$. Lower hydrates tend to retain octahedral coordination by having both waters and anions coordinated to the cation, as for example in $FeCl_2.4H_2O$, which is $[Fe(H_2O)_4Cl_2]$.

15.4.3 Aqua-cations

Salts of transition metal cations generally dissolve in water to give hydrated aqua-cations $[M(H_2O)_x]^{n+}$ (n = 2 or 3; x = 6, except for Pd^{2+} and Pt^{2+}, where x = 4) and the respective anions. The $[M(H_2O)_x]^{3+}$ cations exhibit significant acidity, the electron-withdrawing power of the central M^{3+} cation encouraging proton loss from a coordinated water molecule to give small but significant amounts of the hydroxo complexes $[M(OH)(H_2O)_5]^{2+}$. Whereas in the s- and p-blocks the pK_a values corresponding to such proton loss generally correlate with the charge/sizes of the cations, the pK_a values of the first-row d-block aqua-cations (and indeed also of $Tl^{3+}aq$) correlate with the electron-withdrawing powers of the central cations as reflected in their reduction potentials (Fig. 15.6).

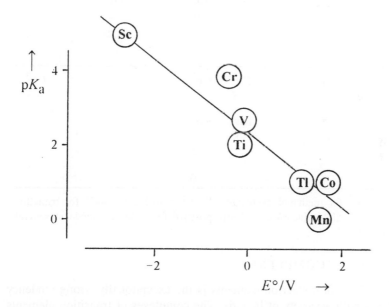

Fig. 15.6   Correlation between pK_a values for $M^{3+}aq$ and reduction potentials for the respective 3+/2+ (3+/1+ for thallium) couples.

On going to higher oxidation states, for all three rows, the increased electron-withdrawing effects of the small highly charged metal centres result in the loss of further protons from coordinated water, leading first to oxo-aqua-cations such as VO^{2+}aq, then to oxoanions such as CrO_4^{2-}, WO_4^{2-}, and MnO_4^-, where the oxide ions may be considered as fully-deprotonated water ligands.

Activation volumes for water exchange reactions of aqua-cations (Table 15.2) afford valuable insight into the pattern of substitution mechanisms for transition metal centres. The trend from negative to positive on going from left to right across the first row indicates a change from associative to dissociative substitution, as the steadily increasing d-electron density makes the approach of incoming water increasingly more difficult. Whereas electronic factors determine the trend in mechanism across the d-block, it is size which appears to determine mechanism going down the Periodic Table. Here, as for the p-block (see e.g. Section 9.3), the formation of an associative transition state of increased coordination number becomes easier as the size of the central metal ion increases.

Table 15.2 – Activation volumes ($\Delta V^{\ddagger}$ / cm^3 mol^{-1}) for water exchange at d-block aqua-cations.

d^0	d^1	d^2	d^3	d^4	d^5	d^6	d^7	d^8	d^9	d^{10}
			V^{2+} -4		Mn^{2+} -5	Fe^{2+} -4	Co^{2+} $+6$	Ni^{2+} $+7$	Cu^{2+} a	
					Ru^{2+} 0			Pd^{2+} -2		
								Pt^{2+} -5		
Sc^{3+} b	Ti^{3+} -12	V^{3+} -9	Cr^{3+} -9		Fe^{3+} -5	Co^{3+} c				Ga^{3+} $+5$
					Ru^{3+} -8	Rh^{3+} -4				
					Ir^{3+} -6					

^{a} $\Delta V^{\ddagger}$ = +8 cm^3 mol^{-1} for methanol exchange. ^{b} $\Delta V^{\ddagger}$ = −12 cm^3 mol^{-1} for trimethyl phosphate exchange. ^{c} $\Delta V^{\ddagger}$ values are modestly positive for water exchange at several ternary Co^{III} complexes.

15.5 FORMATION OF COMPLEXES

A characteristic feature of d electron elements is the exceptionally strong tendency to form complexes with a variety of ligands. The complexes of transition elements can be broadly divided into two groups: (1) classical complexes and (2) non-classical complexes. The first group comprises complexes in which the oxidation number of the central metal atom is well defined and bonding is of the σ type with considerable ionic character. Complexes of this type are formed by halide ions and by ligands containing highly electronegative nitrogen and oxygen atoms as e.g.

ammonia and many polydentate ligands. To the second group belong complexes in which metal–ligand bonding is highly covalent and multiple, because a non-classical ligand can act as both electron donor and electron acceptor. Examples are complexes with such ligands as phosphines, cyanide, or alkenes, and carbonyl and nitrosyl compounds. Due to the double role of the ligand it may be difficult, if not entirely impossible, to assign an unambiguous oxidation number to the central atom. Non-classical complexes merge into organometallic chemistry with no clear dividing line. In addition, complexes of such ligands as 2,2'-bipyridyl link the classical and non-classical groups, their character depending on the nature and oxidation state of the central metal. These matters are considered further in Section 15.5.2 below.

15.5.1 Classical complexes

There are two main factors which favour formation of strong classical complexes by d electron elements. The first is the relatively small ionic radius which for the very common +3 and +4 oxidation states is usually less than 70 pm. The second factor is easy availability of the d orbitals which is a necessary condition for d^2sp^3 hybridization i.e. for formation of octahedral complexes (CN 6). The availability of d orbitals decreases in each series from left to right, see Fig. 2.3 and Table 15.1. According to the data in Table 15.1 the difference between ε_{4s} and ε_{3d} increases from 3.4 V for Sc to 10.8 V for Ni. This increase may be one of the reasons why ions of elements at the end of each transition series, in particular Ni^{2+}, Pd^{2+} and Pt^{2+}, tend to display square-planar coordination (CN 4). This is because for this coordination (hybridization dsp^2) participation of only one d-orbital, the $d_{x^2-y^2}$ orbital, is required.

Both octahedral and tetrahedral complexes of transition elements differ significantly from those formed by p block elements. One of the reasons is splitting of d orbitals in the electrostatic field of the ligands into two sets: t_{2g} and e_g for octahedral, t_2 and e for tetrahedral, complexes, Fig. 15.7. Splitting of the d orbitals in the ligand environment leads to formation of two types of complexes : high-

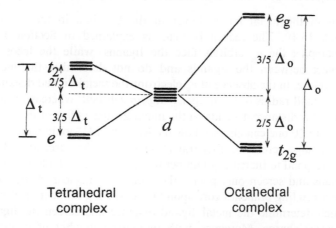

Tetrahedral
complex

Octahedral
complex

Fig. 15.7 Splitting of d orbitals in tetrahedral and octahedral fields.

spin (HS) and low-spin (LS). For instance, Mn(II) in octahedral surroundings forms high-spin complexes, e.g. $Mn(H_2O)_6^{2+}$ (configuration $t_{2g}^3 e_g^2$) and low-spin complexes, e.g. $Mn(CN)_6^{4-}$ (configuration t_{2g}^5). For octahedral complexes both forms may occur when the number of d electrons in the central ion is between 4 and 7, which is the case in the ranges Cr(II)→Co(II) and Mn(III)→Ni(III). Whether LS or HS complexes are formed depends on competition between the exchange energy (see Section 15.6) and the strength of the ligand field. The first factor favours configurations with unpaired spins, while larger splitting of d orbitals in the electrostatic field of the ligands favours configurations with paired spins, Fig. 15.8. The splitting of d orbitals in the ligand environment is, apart from the increasing difference between s and d orbital energies, an important factor which makes coordination numbers decrease across the row. The reason is that at the end of the row electrons must fill either the e_g (octahedral field) or t_2 (tetrahedral field) orbitals. Since the energy gap between e_g and t_{2g} orbitals is much larger than between t_2 and e orbitals (Fig. 15.7), tetrahedral and, in general, less-than-six coordination is preferred at the end of each transition series.

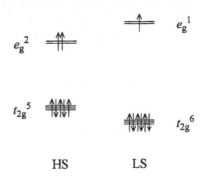

HS   LS

Fig. 15.8 Filling of energy levels in high- and low-spin octahedral Co(II) complexes.

The radius of the metal ion is lower in the LS than in the HS complexes, Figs. 15.9 and 15.10. The reason is that, as explained in Section 15.3, in an octahedral complex the e_g orbitals face the ligands, while the lobes of the t_{2g} orbitals project between the ligands and do not determine the metal–ligand distance. Therefore, in the absence of e_g electrons the metal–ligand distance depends mostly on the small radius of the deeper lying np subshell, whereas in the presence of e_g electrons this distance depends on the much greater radius of the e_g orbitals. In Figs. 15.9 and 15.10 one can also see how the M^{2+} and M^{3+} ionic radii in octahedral coordination change across the first transition series. The descending branches in both plots correspond to increasing number of t_{2g} electrons, which do not determine the ionic radius and screen only poorly the increasing nuclear charge seen by the ligands. The ascending branches correspond to increasing number of electrons in e_g orbitals, which determine the metal–ligand distance and screen the ligands better from the nuclear charge. Moreover, with increasing number of e_g electrons the

repulsion between them increases. This makes their orbital radius, and hence also
the central-ion-to-ligand distance, greater.

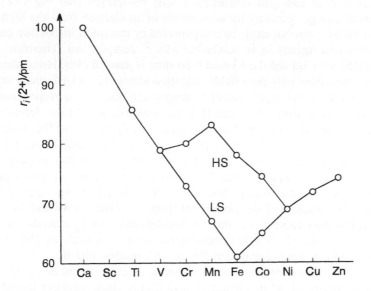

Fig. 15.9 Radii of M^{2+} cations in HS and LS complexes of $3d$ elements and Zn.

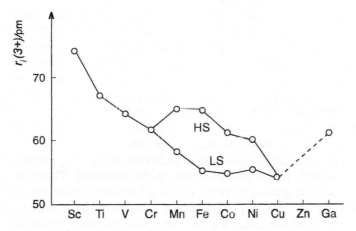

Fig. 15.10 Radii of M^{3+} cations in HS and LS complexes of $3d$ elements and Ga.

How occupancy of d orbitals, split in an octahedral environment, may destabilize
certain oxidation states of d elements is illustrated by the easy oxidation of Co(II) in
low-spin complexes, where Co(II) is in the $t_{2g}^{6}e_{g}^{1}$ configuration, Fig. 15.8. Because
in low-spin complexes the energy gap between t_{2g} and e_{g} orbitals is large (which is
the main reason for formation of a low-spin complex) it becomes easy to detach the
electron from the high lying e_{g} orbital. Another example is the instability of

oxidation state +3 for Pd and Pt. Since most M^{3+} ions of transition elements favour CN 6 one can assume that coordination in hypothetical classical Pt(III) and Pd(III) complexes would also be octahedral with the low-spin electron configuration $t_{2g}^{6}e_{g}^{1}$, the same as that of low-spin octahedral Co(II) complexes (see Fig. 15.8). The relatively small energy necessary for detachment of an electron from the high-lying e_{g} orbital in the M^{3+} ion can easily be compensated by increased interaction between the M^{4+} cation and ligands in an octahedral MX_{6}^{2-} complex ion. Therefore, when Pd(II) and Pt(II) are oxidized the +3 oxidation state is jumped over. Hence these two elements as a rule show only even stable oxidation states (+2 and +4, and 0 in some phosphine complexes and organometallic compounds) and in this respect resemble Group 14 elements. It should be noted that the similarity to p block elements goes even deeper, because the very few known complexes of platinum in the "forbidden" +3 oxidation state show, in addition to bridging acetate ions, metal-to-metal bonds as for example in the complex $[Pt_{2}(\mu\text{-OAc})_{4}(H_{2}O)_{2}]^{2+}$. In this respect Pt(III) resembles e.g. gallium, which in the forbidden oxidation state +2 forms dimeric complexes with metal-to-metal bonds as for instance in the complex $[Cl_{3}Ga\text{–}GaCl_{3}]^{2-}$. Because of the smaller splitting of $3d$ than of $4d$ and $5d$ orbitals in an octahedral environment, i.e. smaller destabilization of e_{g} orbitals, Ni shows the oxidation number +3 in a few monomeric complexes such as $[NiF_{6}]^{3-}$ and $[Ni(bipy)_{3}]^{3+}$, and in a number of complexes with macrocyclic nitrogen ligands – $[Ni(cyclam)]^{3+}$ is even stable in aqueous solution. Nickel also occasionally shows the oxidation number +4, as in complexes with highly electronegative ligands such as fluorine (e.g. in the K^{+} and Ba^{2+} salts of $[NiF_{6}]^{2-}$) or with certain multidentate anionic oxygen/nitrogen donor ligands. The intense colours of complexes of the latter group of ligands indicate significant stabilization through charge-transfer interactions. The reluctance of nickel to show the oxidation number +4 should be contrasted with the stability of the heavier congeners Pd(IV) and, especially, Pt(IV).

The presence of unpaired electrons in t_{2g} or e_{g} orbitals confers paramagnetic properties on many d-ions, with magnetic moments permitting distinction between high- and low-spin complexes. Paramagnetism also permits the observation of electron spin resonance (ESR – or EPR, i.e. electron paramagnetic resonance) spectra, which can provide structural and kinetic information complementary to that obtainable from NMR spectra of diamagnetic species.

The magnitude of d orbital splittings is such that the $t_{2g} \rightarrow e_{g}$ energy difference generally corresponds to the visible region of the spectrum. Therefore ions and complexes of many transition elements (d^{1} to d^{9} configurations) are coloured, in contrast to analogous s and p block ions and complexes. Such colours are weak, since $d \rightarrow d$ transitions are forbidden, but, alongside magnetic properties, were of great importance in the development of Crystal Field Theory. Intense colours in transition metal, and indeed also in some p block complexes, are usually attributable to metal-to-ligand or ligand-to-metal charge-transfer, a process which is always allowed.

Several properties of coordination complexes of the d elements, particularly of the first row transition elements, show an irregular but constant variation across the series. The observed patterns can be attributed to the variations in Crystal Field stabilization on going from d^1 to d^{10} configurations. This may be illustrated for some properties of high-spin 2+ ions. Firstly Fig. 15.11 deals with the special case of hexa-aqua-complexes, showing the d^n electron configurations and Crystal Field Stabilization Energies (CFSE) for the relevant cations and a plot of hydration enthalpy versus d^n. Next Fig. 15.12 shows the dependence of stability constants for complexes of three common ligands on d^n configuration (cf. the long-established Irving-Williams stability sequence $Mn^{2+} < Fe^{2+} < Co^{2+} < Ni^{2+} < Cu^{2+} > Zn^{2+}$).

Cation	Cr^{2+} d^4	Mn^{2+} d^5	Fe^{2+} d^6	Co^{2+} d^7	Ni^{2+} d^8	Cu^{2+} d^9	Zn^{2+} d^{10}
CFSE	$0.6\Delta_o$	0	$0.4\Delta_o$	$0.8\Delta_o$	$1.2\Delta_o$	$0.6\Delta_o$	0
$\Delta H_{hydr}(kJ\,mol^{-1})$	-1850	-1845	-1920	-2054	-2106	-2100	-2044

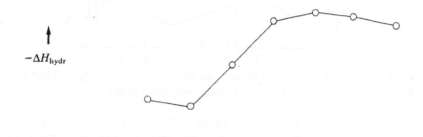

Fig. 15.11 The reflection of CFSE in hydration enthalpies of 2+ cations of first-row d elements.

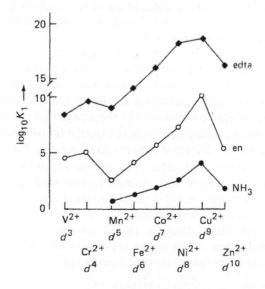

Fig. 15.12 The dependence of stability constants on d electron configuration for high-spin complexes of 2+ cations of first-row d elements.

Here one can see the additional Crystal Field stabilization for the relevant cations in comparison with the d^5 and d^{10} cations where CFSE = 0. Thirdly, Fig. 15.13 illustrates the correlation between reactivity, in the shape of rate constants for water exchange, and CFSE.

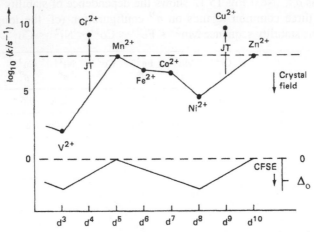

Fig. 15.13 The dependence of rate constants for water exchange on Crystal Field (modified by Jahn-Teller, JT) effects for 2+ aqua-cations of first-row d elements.

In accordance with their high CFSE values, substitution at d^3, low-spin d^6, and d^8 centres is particularly slow. This fact has had an enormous effect on the development of coordination chemistry, for it means that complexes of such cations are generally sufficiently inert for their isolation, purification and characteristion by classical techniques. Thus the first decades of coordination chemistry were dedicated to complexes of chromium(III), cobalt(III) and platinum(II), with the coordination chemistry of other transition metal cations being developed later and more slowly. Substitution timescales were also of great significance in relation to the establishment of electron transfer mechanisms of coordination complexes. Taube's unequivocal demonstration of inner-sphere electron transfer through a transient binuclear Cr–Cl–Co intermediate relies on a combination of water lability in the Cr^{2+}aq reductant and substitution-inertness of the transferred chloride in the $CrCl^+$ product. Complementarily, simple outer-sphere electron transfer between complexes can only be assumed with confidence if both reactants are substitution-inert.

Redox potentials are significantly affected by complex formation, as illustrated for the particular case of iron in Table 15.3. Anionic ligands tend to stabilize the higher oxidation state, ligands which can effect metal-to-ligand charge transfer through π-bonding stabilize the lower oxidation state. The first three entries in Table 15.3 illustrate the potentially large effects of ligand substituents. In the case of e.g. cyanide the two stabilization effects are in opposition and the resultant reduction potential is only a little lower than that of the aqua-couple.

Table 15.3 – Effects of complex formation on reduction potentials
($E°/V$) for the iron(III)/iron(II) couple, in aqueous solution.

Couple [a]	$E°/V$	
$Fe(5\text{-}NO_2phen)_3^{3+/2+}$	+1.25	↑
$Fe(phen)_3^{3+/2+}$	+1.06	Fe(II) stabilized
$Fe(4,7\text{-}Me_2phen)_3^{3+/2+}$	+0.88	│
$Fe(H_2O)_6^{3+/2+}$	+0.77	
$Fe(CN)_6^{3-/4-}$	+0.36	
$Fe(edta)^{1-/2-}$	−0.12	│
$Fe(ox)_3^{0/1-}$	−0.20	Fe(III) stabilized
$Fe_4S_4(SR)_4^{2-/3-}$	−0.58 to −0.63	↓

[a] phen = 1,10-phenanthroline; edta = ethane-1,2-diaminetetra-
acetate; ox = oxinate (8-hydroxyquinolinate); R = aryl.

15.5.2 Non-classical complexes and organometallic compounds

Those transition elements with electronegativities about 2 readily form complexes
with ligands containing donor atoms of similar electronegativity, the most important
of which is the carbon atom. In most of these complexes the ligand both donates a
lone pair to the central atom and accepts electrons from the d orbitals into its empty
antibonding π orbitals. Therefore, these complexes are frequently called complexes
with π-acceptor ligands. The highly electronegative nitrogen atom can also act in
this way, but most frequently when present in molecules with delocalized π orbitals
as e.g. in pyridine, 2,2'-bipyridyl, and 1,10-phenanthroline.

The stoichiometry of non-classical complexes can frequently be predicted from the
18-electron rule which is a counterpart to the octet rule valid for p block elements.
According to the 18-electron rule the total number of valence electrons on the
central atom should be equal to 18. Examples include the $Ni(CO)_4$, $Fe(CO)_5$ and
$Cr(CO)_6$ molecules, in which the total electron count is 10, 8, or 6 electrons from
the metal atom and 8, 10, and 12 electrons from the CO molecules respectively, in
each case totalling 18. The nitric oxide molecule is a three-electron donor – the
electron count for $Co(CO)_3(NO)$ is thus 7 (Co) plus 3 x 2 (3 CO) plus 3 (NO), again
giving 18 electrons around the Co. $V(CO)_6$ is an exception, the V having only 17
electrons, but this compound is keen to add an electron to form the $[V(CO)_6]^-$ anion.
Mononuclear carbonyls of Co and Mn are not stable species – the dimers $Co_2(CO)_8$
and $Mn_2(CO)_{10}$ give the required 18 electron complement to each metal.

As already pointed out, the oxidation number of the central metal atom in non-
classical complexes is not well defined, since the ligands both donate and accept
electrons. In some circumstances it seems more appropriate to consider a ligand as
the radical corresponding to its one-electron reduced form. Thus, for example, some
2,2-bipyridyl complexes behave more like an $M^{(n+1)+}$ complex of bipy$^{\bullet-}$ than an
M^{n+} complex of bipy itself. There are similar uncertainties in the chemistry of many

metal complexes of bidentate aromatic or unsaturated organic ligands with sulphur donor atoms, e.g. maleo-nitriledithiolate, $NC(S^-)C=C(S^-)CN$. On the whole formal oxidation numbers in non-classical complexes, especially those involving carbon-donor ligands, are low (+1, 0 or even negative – as in, e.g., the above $[V(CO)_6]^-$ anion, and in $[Fe(CO)_4]^-$, $[Nb(CO)]_5{}^{2-}$, and $[Cr(CO)_4]^{3-}$). In the organometallic compounds of transition elements, also called π-complexes, both donation and acceptance of electrons is accomplished by the filled bonding and vacant antibonding π orbitals of the ligand. For some of these compounds the 18-electron rule also holds. An example is ferrocene, $(C_5H_5)_2Fe$, in which the two cyclopentadienyl ligands supply 10 electrons and the iron atom 8 electrons.

Cyanide complexes (CN^- is isoelectronic with CO) provide a link between classical coordination complexes and non-classical organo-metallic compounds. The anionic nature of cyanide stabilizes high oxidation states, giving classical coordination complexes such as $[Fe(CN)_6]^{3-}$, $[Mo(CN)_8]^{4-}$ and $[W(CN)_8]^{3-}$. On the other hand cyanide's π-bonding abilities enable it to stabilize certain transition metals in oxidation states as low as zero, for example in $K_6[Cr(CN)_6]$ and $K_4[Ni(CN)_4]$ – cf. $Cr(CO)_6$ and $Ni(CO)_4$ above.

15.6 DIFFERENCES BETWEEN THE SERIES

Homologues in each Group of transition elements are, in general, similar to one another but they may also show remarkable differences, most of all between the first and the next transition series.

15.6.1 Electron configurations of atoms

Atoms of 3d elements show the $3d^n4s^2$ configuration, except for Cr and Cu, which have the $3d^54s^1$ and $3d^{10}4s^1$ configurations respectively. Most atoms of 4d elements have the $4d^{n+1}5s^1$ configuration – Pd even has the $4d^{10}$ configuration – while only Y, Zr and Tc have two electrons in the 5s shell. Since the number of electrons in the outer s shell decreases from 2 to 1 between the first and second series, one might expect this trend to continue, leading to $5d^{n+2}6s^0$ (except for the last members) or at least $5d^{n+1}6s^1$ configurations for 5d elements. Unexpectedly, atoms of the 5d elements revert to having two electrons in their outer s shell, i.e. $5d^n6s^2$ configurations (except for Pt and Au). The electron configuration of transition element atoms is a result of interplay between a number of factors. It has been explained in Section 2.3 why scandium has the $3d^14s^2$ instead of the $3d^24s^1$ configuration, in spite of 3d being below 4s. The main reason is strong interelectronic repulsion in the 3d subshell which has small radius. The argument for the d^ns^2 configuration of Sc can be extended to other 3d elements. The much greater radius of the 4d than the 3d subshell (see Fig. 15.14) and small difference between the radii of the 5s and 4s shells (only 13 pm for the Zr/Ti pair) favour the $4d^{n+1}5s^1$ configuration for the second transition series. The reason is that adding one electron to the large 4d subshell does not increase interelectronic repulsion energy very much. This is effectively overcompensated by the lack of repulsion in the 5s subshell.

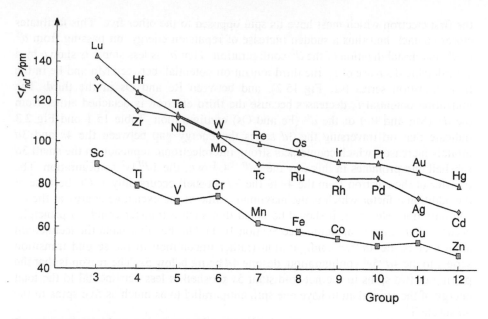

Fig. 15.14 Radii of $3d$, $4d$ and $5d$ orbitals. The $\langle r_{nd} \rangle$ radii are average values weighted over the occupancy of the $d_{3/2}$ and $d_{5/2}$ orbitals. To show the effect of the $4f$ shell in Group 3 Lu has been substituted for La.

Two factors are responsible for the $5d^n6s^2$ configuration of most atoms of the third transition series. The first is the presence of the filled inner $4f$ subshell, which shields the outer $5d$ and $6s$ electrons from the nuclear attraction. This shielding is less effective for $6s$ electrons, which penetrate the atom core deeply, than for $5d$ electrons, which have low probability density near the nucleus (see eq. 1.11). Evidently, incomplete shielding increases stability of the $6s$ orbital more than stability of the $5d$ orbitals and favours the s^2 configuration. An additional factor is the direct relativistic effect, which stabilizes $6s$ electrons.

As already pointed out, Cr has the $3d^54s^1$ configuration whereas the remaining $3d$ elements (except Cu) have two electrons in the $4s$ subshell. On the other hand Tc has the $4d^55s^2$ configuration, whereas its two neighbours to the left (Nb and Mo) and to the right (Ru and Rh) have only one electron in the $5s$ subshell. In both cases the d subshell is half-filled, because the electron is transferred either from s to d (Cr) or from d to s subshell (Tc). In accordance with the Pauli exclusion principle electrons with parallel spins avoid each other, hence repulsion between them is less than if the spins were antiparallel. The decrease in the repulsion energy is proportional to the number of pairs of electrons with parallel spins and is called exchange energy. Because lower repulsion means greater stability of the ground state of the atom, electrons which fill the d subshell in succession have parallel spins up to the d^5 configuration i.e. up to the half-filled d subshell. At this configuration the exchange energy and the respective ionization potential attain their maximum values in the first half of the series. In this sense one can say that the half-filled shell shows unique properties or special stability. On the other hand the sixth d electron is the first electron which must have its spin opposed to the other five. This facilitates closer contact, and thus a sudden increase of repulsion energy, on passing from d^5 to d^6 and destabilization of the d^6 configuration. That d^6 is less stable is shown by a considerable decrease of I_3, the third ionization potential, between Mn and Fe in the first transition series (see Fig. 15.2), and between Re and Os in the third. The ionization potential I_3 decreases because the third electron is detached either from the d^5 (Mn and Re) or the d^6 (Fe and Os) configuration. Table 15.1 and Fig. 2.3 indicate that on traversing the $3d$ series the energy gap between the $4s$ and $3d$ orbitals increases, which counteracts strong interelectronic repulsion in the small $3d$ subshell and promotes in this way the $3d^{n+1}4s^1$ over the $3d^n4s^2$ configuration. The transfer of the electron from the $4s$ to the $3d$ subshell occurs only at Cr, because of the additional factor which is the maximum value of the exchange energy at the d^5 configuration. However, it should be noted that such a transfer could, in principle, occur earlier, e.g. at vanadium (see Section 16.1). On the other hand the technetium atom prefers the $4d^55s^2$ configuration (rather uncommon in the second transition series) to the $4d^65s^1$ configuration, despite $4d$ being below $5s$. The reason is that the pairing of two spins in the not-so-distant $5s$ subshell is less detrimental to the total energy of the atom than to have one spin antiparallel to as much as five spins in the $4d$ subshell.

15.6.2 Oxidation states

Higher oxidation states are as a rule more stable for the $4d$ and $5d$ than for the $3d$ elements. For instance, Mo(VI) and W(VI), and also Tc(VII) and Re(VII), form stable MO_4^{2-} or MO_4^- oxoanions respectively. On the other hand the Cr(VI) and Mn(VII) oxoanions CrO_4^{2-} and MnO_4^- are strong oxidants. Os and Ru form tetroxides, where the formal oxidation state is +8, while the highest oxidation state for Fe is +6, in FeO_4^{2-}. Cobalt is stable as Co(II) (in high-spin complexes), whereas the characteristic oxidation numbers of Rh and Ir are +3 and +4. Similarly, Mn(II) is stable in many ligand environments, while Tc and Re only very rarely show the oxidation number +2. The main reason is that the ionization potentials I_3 and I_4 are smaller for $4d$ and $5d$ than for $3d$ elements (see Table 15.4). Smaller ionization potentials mean, of course, less expenditure of energy to attain higher oxidation states.

It is seen in Table 15.4 that changes in ionization potentials parallel changes in the absolute values of the nd orbital energies, while radii show the opposite trend. The small radius of the $3d$ shell arises from the lack of inner d orbitals, while repulsion between the $3d$ and $4d$ orbitals makes the radius of the $4d$ shell quite large and ε_{4d} less negative. This effect is illustrated in Fig. 15.15, which shows the small gap in radial extent between the $3d_{3/2}$ and $3p$ orbitals in Sc and much greater gaps between the corresponding orbitals in Y, Lu and Ac. Fig. 15.15 should be compared with Fig. 3.3 which shows similar effects for p orbitals. The proximity of $4d_{3/2}$ and $5d_{3/2}$ (in general $4d$ and $5d$) orbitals, as also $4p$ and $5p$, with regard to radius and energy arises from the presence of the filled $4f$ shell which shields the $5d$ electrons incompletely from the nuclear charge. The effect of the $4f$ shell is similar to that of the $3d$ shell on $4p$ electrons (see Section 3.2).

Table 15.4 – Orbital energies, ionization potentials and orbital
radii of Group 7 elements

	Mn	Tc	Re
ε_{nd} / eV	−15.0	−12.4	−10.6
I_3 / eV	33.7	29.5	26.0
I_4 / eV	51.2	42.5	37.7
$\langle r_{nd} \rangle$ / pm	61.5	88.2	97.5

It is known that higher oxidation states of Group 9 and 10 elements are more stable for $5d$ than $4d$ elements, respectively. For instance, the +4 oxidation state is more stable for Ir and Pt than for Rh and Pd. The main reason is the significantly smaller fourth ionization potential of the heaviest member in each of the two Groups. I_4 is about 39 eV for Ir and 45 eV for Rh, about 40 and 49 eV for Pt and Pd, respectively. Because for all the elements of Group 9 and 10 the fourth electron is detached from the $d_{5/2}$ subshell, one may presume that in addition to the general trend of decreasing potential down each Group, the decrease in the fourth ionization potential results from greater relativistic destabilization of the $5d_{5/2}$ than of the $4d_{5/2}$ orbitals. Probably for the same reason, strong destabilization of the $d_{5/2}$ orbital, only Pt in Group 10 (configuration $5d_{3/2}{}^4 5d_{5/2}{}^5 6s^1$) shows the oxidation number +6 (in the hexafluoride), but in this oxidation state is a very strong oxidant able to oxidize Xe to Xe^{2+}.

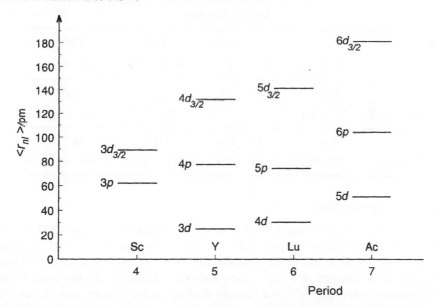

Fig. 15.15 The difference in radial extent between the $nd_{3/2}$ orbitals and np subshells in Group 3 elements. The radii of closed p subshells are average values weighted over the occupancy of $p_{1/2}$ and $p_{3/2}$ orbitals.

15.6.3 Complexes

Complexes of $4d$ and $5d$ elements are as a rule low-spin, while those of $3d$ elements can be either low- or high-spin, depending on the position of the ligand in the spectrochemical series. The main reason is the greater radial extent of $4d$ and $5d$ orbitals, which makes them more sensitive toward the ligand field than $3d$ orbitals and results in greater splitting. Greater splitting means greater expense in energy is necessary to locate an electron in the e_g orbital. The ionic radii of homologues in a Group are lower for $3d$ than for $4d$ and $5d$ elements, while the latter have very similar ionic radii, Table 15.5. As already pointed out and shown in Figs. 15.3 and 15.4, the radii of the 3+ and 4+ ions with empty e_g orbitals depend on the radius of the deeper lying np shell. The radius $\langle r_{np} \rangle$, like the radius $\langle r_{nd} \rangle$, increases between the first and second transition series and becomes almost the same for the second and third series. Table 15.5 shows, as an example, that changes in r_i down Groups 4, 5, 6 and 9 do indeed parallel changes in $\langle r_{np} \rangle$. The smallness of the change in the radius $\langle r_{np} \rangle$ between homologues in the second and third transition series is due to the presence of the filled $4f$ shell which incompletely shields the more distant $5p$ electrons from the nuclear charge. The characteristic coordination number of transition element ions is 6 and 4 (4 at the end of each series). However, ions of $4d$ and $5d$ elements at the beginning of each series show also CN 7 and 8, as e.g. in $Mo(CN)_7^{4-}$, $Mo(CN)_8^{3-,4-}$ and $[Zr, Hf(acac)_4]$ (acac denotes the acetylacetonate, alias 2,4-pentanedionate, ligand). The reason is that the ionic radius is sufficiently large to accommodate seven or even eight ligands around the central metal ion.

Table 15.5 – Ionic (CN 6) and np orbital radii (pm) of some d block elements

	V^{3+}	Co^{3+}(LS)	Ti^{4+}	Cr^{4+}
r_i, $\langle r_{np} \rangle$	64.0 54	54.5 43.3	60.5 57.8	55 51.4
	Nb^{3+}	Rh^{3+}(LS)	Zr^{4+}	Mo^{4+}
r_i, $\langle r_{np} \rangle$	72, 69.4	66.5, 57.7	72, 72.7	65, 65.9
	Ta^{3+}	Ir^{3+}(LS)	Hf^{4+}	W^{4+}
r_i, $\langle r_{np} \rangle$	72, 69	68, 60.4	71, 71.5	66, 66.6

15.6.4 Other properties

A prominent feature of Group 6, and to a more limited degree also of Group 5, elements is the formation of polyacids and their salts. In this respect the first element in the Group again differs considerably from its heavier congeners. Thus, chromium in the tetrahedral CrO_4^{2-} anion through sharing one oxygen atom forms the dichromate anion, $Cr_2O_7^{4-}$, and through sharing further oxygen atoms forms the tri- and tetra-chromates which, however, are of very much less importance. On the other hand molybdenum forms, among many other polymetallates, the octamolybdate anion $[Mo_8O_{26}]^{4-}$ and tungsten the highly condensed paratungstate anion, $[W_{12}O_{41}]^{10-}$. However, the difference between chromium and its heavier congeners is not only in the degree of condensation but in different coordination of oxygen atoms around the central metal ion. Thus, Mo and W form polyacids through sharing of oxygen atoms in octahedral MO_6 groups, but chromium in tetrahedral MO_4 groups. That Mo and W show a higher coordination number than Cr in polyoxoacids is the result of the greater radii of the $4p$ and $5p$ shells than of the $3p$ shell; greater radius makes possible accommodation of more oxygen atoms around the central metal ion. It should be remembered that ionic radii of d elements depend on the radius of the outermost p subshell – see Section 15.3. Another factor is probably formation of stronger $p\pi$-$d\pi$ bonds in $CrO_4{}^{2-}$ than in $MoO_4{}^{2-}$ and $WO_4{}^{2-}$. The $p\pi$-$d\pi$ bonds in these oxoanions are formed by overlap of p orbitals on the oxygen atom with d orbitals on the metal atom. Better overlap in the case of the Cr atom results from the smaller metal to oxygen distance in $CrO_4{}^{2-}$, due to smaller $3p$ than $4p$ and $5p$ orbital radii. The presence of $p\pi$-$d\pi$ bonds hinders the increase of coordination number, because such an increase would make the metal to oxygen distance larger and π-bonding weaker. The difference in coordination number between chromium and its homologues in oxoanions bears some resemblance to the difference in catenation between oxygen (which like chromium is the first element in the Group) and sulphur. We know from Section 12.1 that the ease of formation of $p\pi$-$p\pi$ bonds by small oxygen atoms favours formation of O_3 molecules over long chains. The much larger sulphur atoms are reluctant to form such bonds, hence they catenate freely.

The first series metals form few or no M–M bonded species, except in polynuclear carbonyl complexes, whereas cluster formation is common for heavier congeners in low oxidation states. Examples are the niobium and tantalum halide clusters $[M_6X_{12}]^{2+}$, which have no vanadium analogues. The $[M_6X_{12}]^{2+}$ units consist of octahedral clusters of metal ions with bridging halide ions over each edge of the octahedron. In this cluster each M atom is bonded to four M and four X atoms i.e. its formal coordination number is 8. Moreover, in the solid state one additional X^- ligand (bridging two $[M_6X_8]$ units) is coordinated to the metal atom, which makes its CN even higher. Another example is provided by the $[M_6X_8]^{4+}$ clusters of molybdenum and tungsten, but not of chromium. The structure of the $[M_6X_8]$ unit is an octahedron of metal atoms with the bridging halide ion over each triangular face. Here again the metal atom is bonded to four M and four X atoms. The high coordination numbers of metals in clusters are probably the reason why the smaller metal atoms of the first series do not form halide-containing clusters. Another but related reason is the distance between the $\varepsilon_{(n+1)s}$ and ε_{nd} orbital energies, which is much greater for the first member of the Group than for its heavier congeners. The greater energy gap hinders the formation of hybrid orbitals with participation of a large number of d orbitals which is required by high coordination numbers.

It should be noted that in the $[M_6X_{12}]^{2+}$ and $[M_6Cl_8]^{4+}$ clusters the oxidation numbers of the metal atoms are +2.33 for Group 5 and +2 for Group 6 elements. Therefore, clusters of this kind present exceptions to the general rule, according to which lower oxidation states are more stable in the elements of the $3d$ than the $4d$ and $5d$ series. These unusually low oxidation states, as for intermediate elements in the second and third d series, are stabilized by formation of M–M bonds, in addition to M–X bonds.

The unique feature of Mo, W, Tc and Re is the formation of metal to metal multiple bonds which is not shown, in principle, by their lighter $3d$ congeners Cr and Mn. The probable reason is the much greater radial extent of $4d$ and $5d$ than of $3d$ orbitals, which results in more efficient orbital overlap. Admittedly, Cr in oxidation state +2 also forms quadruple metal to metal bonds as e.g. in the numerous acetates of the general formula $Cr_2(O_2CCH_3)_4L_2$, where L is a donor ligand. However, their structure is typical of carboxylato-bridged dinuclear complexes where the bridging acetato groups keep the two chromium atoms near to each other, thus facilitating formation of the quadruple bond.

15.6.5 Comparison with p block elements

The d and p block elements differ in almost all properties. Examples include colour and magnetic properties of ions, stability of complexes, and different oxidation state patterns. Some of these differences have already been explained by the accessibility and ligand field splitting of d orbitals, and the small decrease of the ligand-to-metal distance on detaching a d electron. However, the most striking difference is that all transition elements are metals, whereas in the p block only the Group 13 elements (except B), β-Sn, Pb, As, Sb, Bi, and Po have metallic properties. The reason is that transition elements have two bands – the s/p and the d band. Therefore, as a transition series is traversed electrons fill in succession bonding and antibonding orbitals in the d band, whereas in the s/p band even the bonding orbitals remain only partially filled. This makes possible formation of a metallic phase even at the end of the series, albeit less strongly bonded than around the middle. In contrast, most solid p block elements employing their s and p orbitals form a filled bonding, a filled nonbonding and a higher-lying antibonding band which remains empty across the row, see Section 6.5. This pattern of band filling precludes the movement of electrons under an applied electric field, hence their lack of metallic properties. From the chemical point of view the filled bonding band corresponds with covalent bonds between the atoms and the nonbonding band with the lone electron pairs.

16

Group 11. The coinage metals

Table 16.1 – Fundamental properties

	Cu	Ag	Au
$R = \langle r_{ns} \rangle$ / pm	173	183	162
I_1/eV	7.72	7.58	9.22
I_2/eV	20.3	21.5	20.5
I_3/eV	36.8	34.8	30.0
A/eV	1.23	1.30	2.31
χ	1.90	1.93	2.54
mp (°C)	1083	962	1064
r_{met} / pm	128	144	144
$r_i(1+)$ / pm (CN 2)	49	68	62
$r_i(2+)$ / pm (CN 6)	73	94	
$r_i(3+)$ / pm (CN 6)	54	75	85
$E°$ / V	+0.340 (2+/0)	+0.799 (1+/0)	+1.00 (3+/0) [a]
	+0.159 (2+/1+)	+1.980 (2+/1+)	+1.15 (1+/0) [b]

[a] For the reaction $AuCl_4^- + 3e^- \rightarrow Au + 4Cl^-$.   [b] For the reaction $AuCl_2^- + e^- \rightarrow Au + 2Cl^-$.

16.1 GENERAL PROPERTIES

The Group 11 elements, known also as the coinage metals, are d elements according to the broader definition, but because of some remarkable properties they deserve separate discussion.

As most elements in the $4d$ series have only one electron in the $5s$ subshell (zero in the case of Pd) one can expect the configuration $d^{10}s^1$ for the Ag atom, but not for Cu and Au atoms. That all Group 11 elements show the $d^{10}s^1$ configuration is commonly attributed to the stability of the filled shell. We know (Section 15.5) that as one moves from d^1 to d^5 the exchange energy increases, because of increasing number of electron pairs with parallel spins. Simple considerations show that the pattern of increasing exchange energy is repeated from the d^6 to the d^{10}

configuration. At the same time, again as in the first half of the transition series, the gap between the outermost s and d orbitals increases. The two factors, increasing energy gap and increasing exchange energy, promote the transfer of the electron from the ns to the $(n-1)d$ subshell. In the first transition series this transfer occurs at the Cu atom which acquires in this way the $d^{10}s^1$ configuration with its "magic" filled d shell. However, in the third transition series not only gold but also the preceding element – platinum – have only one electron in the $6s$ subshell. It is significant that in the Pt atom the transfer of the electron from $6s$ to $5d$ results not in a filled d shell, but in the d^9s^1 configuration. The example of platinum clearly shows that the filled (and also half-filled) shell effect, i.e. maximum exchange energy, is an important but not the only factor engendering the transfer of the electron when the first half or the end of the transition series is approached.

Despite the fact that elements of Group 11 and Group 1 have only one electron in the outer ns shell they differ radically in respect to most properties, except for showing the oxidation number +1. Thus the first ionization potential of a Group 11 element is much higher than that of the Group 1 element belonging to the same Period, while the opposite relationship holds for their atomic radii. Both effects are due to less efficient screening of the nuclear charge for the s electron by filled d than by filled p shells. In contrast to I_1, the ionization potentials I_2 and I_3 are much lower for Group 11 elements than for the alkali metals. The reason is that the second and third electrons are detached from the lower lying $(n-1)d$ subshell in Group 11 elements, whereas in Group 1 elements they are taken from the very low lying $(n-1)p$ subshell. Therefore, copper, silver and gold display not only the oxidation number +1 characteristic of the alkali metals, but also the oxidation numbers +2 (except Au) and +3. Because of their high first ionization potentials, the Group 11 elements are chemically inert, in contrast to the very reactive Group 1 elements.

The Group 11 elements all crystallize with the same face-centered cubic structure, and in the metallic state are very similar to one another. In comparison with other d elements they have low melting and boiling points and show very high electric conductivity and malleability. Melting and boiling points are low because in Group 11 elements the d band is, in principle, completely filled. One should also note that melting points change in the order Cu > Ag < Au and that the metallic radii follow the sequence Cu < Ag = Au (see Table 16.1). The reversal of the two tendencies between Ag and Au originates in the direct relativistic effect in Au and in the presence of the filled $4f$ subshell. These both decrease the ε_{6s} orbital energy and the $\langle r_{6s} \rangle$ orbital radius thereby making metallic bonding stronger. Atomization energies and melting points of Group 11 elements are much higher than those of Group 1 elements. The reason is that in the coinage metals d orbitals of appropriate symmetry can mix (hybridize) along certain crystallographic directions (particularly along the [110] direction) with s and p orbitals to form a common $s/p/d$ band which contains more than one itinerant electron per atom. The greater number of electrons in the energy band of coinage metals than of alkali metals makes metallic bonding much stronger in the former.

Group 11 elements combine mainly with Group 16 elements and the halogens. Oxides of Ag and Au are much less stable than those of Cu. Selenides and tellurides are metallic, usually non-stoichiometric, some showing superconductivity at low temperature. With halogens only gold forms a pentahalide (with fluorine) and trihalides and, except for AgF_2, only copper forms dihalides. Monohalides are sparingly soluble in water but dissolve in solutions of high halide concentration and in aqueous ammonia. An exception is AgF which forms hydrates and is freely soluble in water. The reasons are its ionic character, which in turn results from the very high electron affinity of fluorine, and the high hydration energy of fluoride.

Monohalides of silver, particularly AgBr, are photosensitive. When exposed to light the halide ion loses its electron to the conduction band through which it reaches the surface of an AgX crystal where it reduces Ag^+ to Ag^0. Gold trihalides, like aluminium trihalides, form dimers, Au_2X_6 (X = Cl, Br), in which Cl or Br atoms act as bridging atoms. This is because sharing of electrons by a halogen atom with two metal atoms requires high electronegativity of the latter, which is shown just by gold. However, in contrast to aluminium the gold dimers are square-planar, the characteristic coordination geometry of Au(III). Au_2Cl_6 and $[AuCl_4]^-$ are the most commonly encountered gold-containing species. They are the usual starting materials for the preparation of Au(III) complexes, usually obtained by ligand substitution from $[AuCl_4]^-$. The $[AuCl_4]^-$ ion is easily reduced to give intensely coloured colloidal solutions of metallic gold.

There are remarkably few stable aqua-ions of the coinage metals. Cu^{2+}aq and Ag^+aq are familiar entities, but Cu^+aq is unstable with respect to disproportionation, Ag^{2+}aq rapidly reduced, in aqueous media. There are no well-established aqua-cations of gold. Copper, silver and gold readily form complexes in each of their oxidation states. Many copper(I) and several silver(II) complexes are stable in aqueous solution, in contrast to the instability of their parent aqua-cations. Many complexes of silver(III), gold(I), and gold(III) are stable in solution as well as in the solid state. This extensive coordination chemistry is in remarkable contrast to the behaviour of Group 1 elements. The first factor which favours formation of complexes by the Group 11 M^+ ions is their much lower hardness i.e. their greater polarizability. For instance, the hardness calculated from eq. 4.11 is 6.28 for Cu^+, while that of Li^+, which has almost the same ionic radius (for CN 6) is 35.1 eV. Greater polarizability results in considerable contribution (particularly in Au) from covalence to interaction between metal s orbital and ligand p orbitals, which makes bonding stronger. Another factor contributing to the much stronger complexing by Group 11 than by Group 1 elements is the less efficient screening from the nuclear charge by the filled d than by the p subshell, which means that the attraction felt by the ligand electrons is distinctly greater in the case of Group 11 than Group 1 cations, hence the increased stability of complexes. Copper(I), silver(I), gold(I) and gold(III) all show relatively little affinity for oxygen donor atoms and prefer formation of complexes with halides and with ligands which contain nitrogen or sulphur donor atoms. The reason is again the high softness of the respective metal ions, which favours formation of partially covalent complexes with soft or at least softer-than-oxygen donor atoms.

The oxidizing powers of gold(I) and gold(III), and the reducing powers of copper(I), impose severe restrictions on which ligands can be used for complex formation at these centres, and indeed on which solvents may be used for studying complex formation in solution. In practice only Cu^{2+} and Ag^+ are suitable for stability constant determinations in aqueous solution. The remarkable feature of copper(II) coordination chemistry is the high stability of these complexes in comparison with their analogues of the other first row transition metal 2+ cations. Steadily rising stability from manganese(II) to copper(II), with a sharp drop to zinc(II), is ubiquitous, as has been generally appreciated since the establishment of the Irving-Williams order half a century ago (see Section 15.5.1, p. 145). Some silver(I) complexes have high stability constants – for example that for formation of the thiosulphate complex is sufficiently favourable for silver chloride to dissolve in an aqueous thiosulphate solution. Similarly, since gold(I) forms the very stable complex $[Au(CN)_2]^-$, gold metal dissolves in cyanide solutions in the presence of aerial oxygen – a key step in the extraction of gold from gold-bearing deposits.

The three metals in their various oxidation states exhibit a variety of stereochemistries. The d^8 centres Cu(III), Ag(III) and Au(III) are expected to adopt square-planar geometry, as indeed the majority of the numerous Au(III) complexes (and, as mentioned above, the Au_2X_6 dimers) do. Such species as $[Ag(OH)_4]^-$ and most of the structurally-characterized Cu(III) complexes are also square-planar. Copper(II), with its d^9 configuration, is prone to undergo Jahn-Teller distortion, usually to give tetragonal species with four short and two long bonds (causing some concern about the precise meaning of the ionic radius for this cation!). Thus, for example, the $[Cu(OH)_6]^{4-}$ anion, in its barium salt, has four O atoms at 197 pm and two at 281 pm. However other forms of distortion are known, for instance $[Cu(H_2O)_6]^{2+}$ has three pairs of oxygens at different distances, these being 209, 216, and 228 pm in its perchlorate salt. The relatively weak attachment of the more distant water molecules in $[Cu(H_2O)_6]^{2+}$ has marked kinetic consequences. Water exchange at this cation is extremely fast, with a rate constant of 5×10^9 s^{-1} (at 298 K) approaching the diffusion-controlled limit. Ligand substitution by simple monodentate ligands is comparably fast, but substitution rates drop dramatically as incoming ligands increase in denticity and in steric demands and constraints. Thus rate constants decrease from the order of 10^9 dm^3 mol^{-1} s^{-1} for ligands such as ammonia or pyridine, through $\sim 10^7$ for open-chain tetradentate amine ligands of the $H_2NCH_2CH_2NHCH_2CH_2NHCH_2CH_2NH_2$ type and $\sim 10^3$ for flexible macrocyclic analogues, to $\sim 10^{-2}$ for rigid porphyrins. This reactivity range of 10^{11} or more is unparalleled for complex formation at transition metal centres. Copper(I) favours tetrahedral stereochemistry, as expected from its d^{10} configuration, but silver(I) and gold(I) generally favour a CN of two, in linear coordination. The differing stereochemical predilections of copper(I) and copper(II) present a potential problem in relation to bioinorganic systems where electron transfer involves Cu(I)/Cu(II) cycling. There could need to be significant stereochemical change, involving both CN and geometry, accompanying electron transfer, which would prevent the very fast electron transfer which is often required in living systems. In practice this problem is often circumvented by the copper in electron-transfer metalloproteins adopting the compromise geometry of flattened tetrahedral (D_{2d} local symmetry).

The radius, favoured tetrahedral configuration, and affinity for nitrogen-donor ligands of copper(I) make it uniquely suited to act as a template for the formation of catenanes and molecular knots. Fig. 16.1(a) shows a typical example of a cyclic phenanthroline–polyether catenand ligand, Fig. 16.1(b) and (c) two such rings inter-linked and bonded to Cu(I) in the catenane complex. Fig. 16.1(d) shows a potential molecular knot assembled on a pair of Cu(I) template ions. The free catenane or knot (shown in outline in Fig. 16.1(e)) can be obtained from their respective Cu(I) complexes simply by demetallation with cyanide.

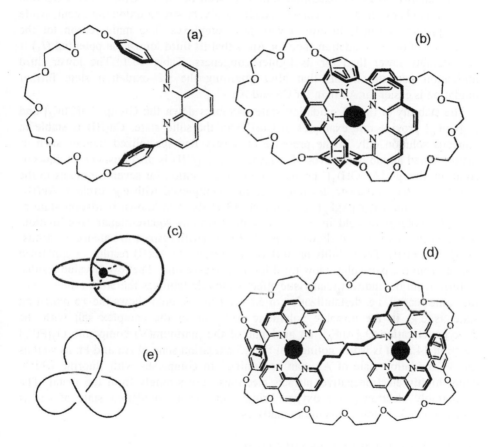

Fig. 16.1 (a) A typical phenanthroline–polyether catenand ring. (b) The Cu(I)–bis-catenand complex of such a ligand. (c) Schematic representation showing the linked rings in (b) and their attachment to the Cu(I). (d) The bis-Cu(I) precursor complex of a molecular knot. (e) A trefoil knot – the idealized form of the ligand in (d).

16.2 CHANGES IN PROPERTIES DOWN THE GROUP

Chemically there is moderate similarity between copper and its heavier congeners silver and gold, which are themselves more alike in some respects. The Group 11 elements differ mainly in the stability of their +1, +2 and +3 oxidation states, as summarised in the following paragraphs.

The oxidation state +1 is characteristic for silver, which in aqueous solutions forms hydrated Ag^+ cations. Cu(I) and Au(I) are not stable as hydrated cations, though solvated Cu^+ is stable in, e.g., acetonitrile solution, and Cu(I) and Au(I) form a number of complex ions which are stable in aqueous solutions.

The stability of the +2 oxidation state decreases down the Group. The oxidation state +2 is characteristic for copper, Ag(II) is a very strong oxidizing agent, while gold appears as Au(II) in only a very few complexes. The main reason for the instability of the +2 oxidation state for Au is that its third ionization potential (I_3) is considerably lower than for its lighter congeners, Table 16.1. The lower third ionization potential means that after attaining the +2 oxidation state further oxidation is easier for Au than for Cu and Ag.

The stability of the +3 oxidation state increases down the Group. $Cs[CuF_4]$ and $K_3[CuF_6]$ provide rare examples of Cu(III) in the solid state; Cu(III) is stable in aqueous solution only in the presence of a very few specialized ligands such as deprotonated amido derivatives of macrocycles. Ag(III) is also unstable in aqueous solution, though $[Ag(OH)_4]^-$ persists in alkaline solution for several minutes in the presence of e.g. carbonate, for longer periods complexed with e.g. cyclam. Ag(III) forms the solid salt $K[AgF_4]$. In contrast, +3 is the best known oxidation state of gold. Complexes of gold in oxidation state +3 show square-planar coordination, which is the same as that shown by isoelectronic (with respect to valence electrons) Pd(II) and Pt(II). The stability of Au(III) with respect to Au(II) results, as explained above, from the particularly low third ionization potential. The low I_3 value results, in turn, from the indirect relativistic effect which destabilizes the $5d_{5/2}$ orbitals. For the same reason, i.e. destabilization of $5d_{5/2}$ electrons, gold shows the +5 oxidation state as e.g. in the unstable polymeric AuF_5 or in the complex salt with the dioxygenyl cation $O_2[AuF_6]$, an analogue of the platinum(V) compound $O_2[PtF_6]$ (Section 14.2). It is worth noting that the +5 oxidation state of Au and Pt, as well as the 3+ oxidation state of Ag, can be attained in complexes with fluorine which, being highly electronegative, attracts electrons very strongly from the metal. The comparably electronegative oxygen brings out the 4+ oxidation state of Cu as transient CuO_2 in low temperature matrices.

16.3 UNIQUE PROPERTIES OF GOLD

The unique properties of gold arise, as already pointed out, from the presence of the filled $4f$ shell and from the direct relativistic effect, which both stabilize the $6s$ orbital. Stabilization of $6s$ electrons makes the first ionization potential, electron affinity and electronegativity of gold anomalously high, and the radius of the gold atom much lower than that of silver, and even lower than that of copper. The indirect relativistic effect, which destabilizes the $d_{5/2}$ orbital in gold, is also an important factor. Because of relativistic and shell effects (the presence of the filled $4f$ subshell) gold differs markedly from its lighter congeners in several respects.

– Gold is much more noble than copper and silver ($6s$ electron stabilized).

– Gold is one of four coloured metals (the others are Cu, Cs and Bi). The red colour of copper is due to light absorption in the green and blue regions. This absorption is assigned to a transition of an electron from the filled d band to the $6s$ Fermi level. The Fermi level is the energy corresponding with the highest filled level in a partly filled band. In the case of silver the respective energy difference corresponds to absorption in the ultraviolet, hence no colour is seen. In the case of gold the gap between the filled $5d$ band and $6s$ Fermi level becomes smaller, which results in light absorption in the range very similar to that of copper (see Table 16.2). The proximity of the s Fermi level and the d band in metallic gold is the result of relativistic stabilization of the $6s$ orbital and destabilization of the $d_{5/2}$ orbitals (Fig. 16.2). Nonrelativistic gold would thus be white, like silver.

Table 16.2 – Energy gap between the $(n-1)d$ band and the ns Fermi level

	Cu	Ag	Au
Energy gap / eV	2.3	4.0	2.4

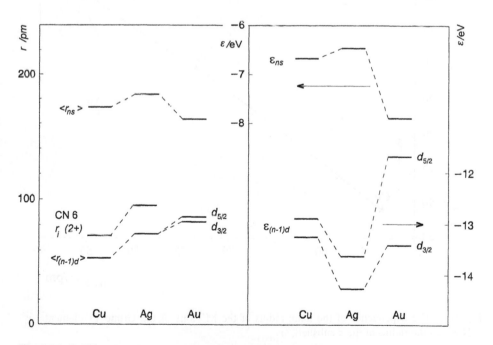

Fig. 16.2 Radii and orbital energies of Group 11 elements.

– Gold forms the salt Cs^+Au^- (which is a semiconductor) and dissolves in solutions of alkali metals in liquid ammonia to give the solvated $[Au(NH_3)_n]^-$ anion. Both properties originate in the exceptionally high electronegativity and electron affinity of gold, which in turn are due to stabilization of the $6s$ orbital.

– The ionic radius of Au(I) for CN 2 is much lower than expected from linear dependence on the radius of the outermost shell in the ion, which is the $(n-1)d$ shell (compare Fig. 16.3 with Figs. 3.6, 15.3 and 15.4). A plausible explanation is again relativistic stabilization of the $6s$ orbital, which makes bonding in gold complexes more covalent than in its lighter congeners. Significant contribution from covalence to bonding makes the Au–X distance in $[MX_2]^-$ complexes shorter. The reason why the effect is so large for CN 2 is the sp hybridization which has the largest possible share of the relativistically stabilized $6s$. Gold in the gaseous phase forms the Au_2 molecule (isoelectronic with the dimeric Hg_2^{2+} ion) which is much more stable than the Cu_2 and Ag_2 molecules. The smaller atomic radius and more negative orbital energy ε_{ns} of the Au atom promote strong σ bonding in the Au_2 molecule. The tendency of gold to form strong Au–Au bonds is also manifested in formation of colloidal forms and clusters.

– Due to relativistic destabilization of the $5d_{5/2}$ orbitals gold, as already pointed out, shows a stable +3 and a modestly stable +5 oxidation state.

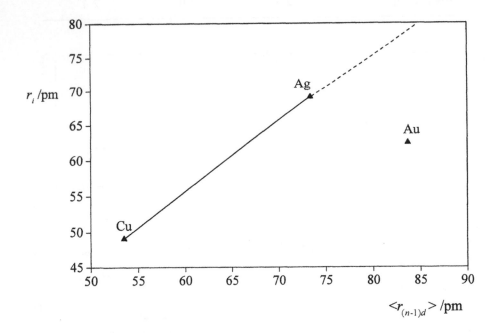

Fig. 16.3 The dependence of the ionic radius of the M^+ ions of the Group 11 elements, r_i (CN 2), on the radius of the d subshell, $\langle r_{(n-1)\,d}\rangle$.

17

Group 12

Table 17.1 — Fundamental properties

	Zn	Cd	Hg
$R = \langle r_{ns} \rangle$ / pm	151	163	150
I_1 / eV	9.39	8.99	10.43
I_2 / eV	18.0	16.9	18.8
I_3 / eV	39.7	37.5	34.2
χ	1.65	1.69	2.00
mp (°C)	420	321	−38.9
bp (°C)	907	765	357
r_{met} / pm	133	149	160
$r_i(2+)$ / pm (CN 6)	74	95	102
$E°(2+/0)$ / V	−0.763	−0.403	+0.853

17.1 GENERAL PROPERTIES

The Group 12 elements are not d-electron elements, neither according to the normal nor to the broader definition. However they do show some similarity to the transition elements, most of all in their tendency to form complexes with ammonia, amines, halide and pseudo-halide (CN⁻, NCS⁻) ions. On the other hand their ability to act as d-π donors is so low that they are extremely reluctant to form other typical transition metal compounds, such as carbonyls, nitrosyls or π-complexes with alkenes. The reason is the high value of the third ionization potential, I_3, at the d^{10} configuration, see Fig. 15.2. A high I_3 discourages partial transfer of a d electron from the metal ion onto the ligand.

With respect to the electron configuration of their atoms the Group 12 elements are s-electron elements, but their similarity to Group 2 elements is limited mainly to their showing the same +2 oxidation state. However, one should note that zinc, cadmium and mercury resemble magnesium in the formation of organometallic compounds RMX and R_2M, where R denotes an alkyl or aryl group and X is a halogen atom. The R_2M molecules are linear, form non-polar liquids or low-melting solids freely soluble in organic solvents. The lipophilic character of dimethyl mercury favours penetration through cell membranes, which makes this compound particularly toxic. R_2Zn and R_2Cd react vigorously with air, oxygen and water. Oxygen and water affect the mercury dialkyls and diaryls much less, presumably because of higher covalency contributions to bonding in the R_2Hg molecules. The highly non-polar character of the Hg–C bond results from the relatively small difference in electronegativity between mercury and carbon. Organomercury compounds find wide application in the preparation of organometallic compounds of other metals by exchange reactions. The number of valence electrons prevents the Group 12 elements, like the Group 2 elements, from forming M_2 molecules.

Zinc, cadmium and mercury, again like Group 2 elements, would not form metallic phases were it not for overlap of the s and p bands. The energy of metallic bonding decreases from zinc to mercury, and is lower than this energy of Group 2 elements in the same row, which is shown by lower atomization energies and melting points of Group 12 than of Group 2 metals (see Tables 8.1, 17.1 and Figs. 6.3, 15.1). This contrasts with what is observed when Group 11 and Group 1 metals are compared (see Section 16.1) and shows that in Group 12 elements the energy band has exclusively s/p character. The d band in these elements has a very narrow energy spread, is well below the s/p band and, because it is filled, does not contribute to metallic bonding. The particularly weak metallic bonding in mercury, which results in a very low enthalpy of atomization and exceptionally low melting and boiling temperatures is most probably due to the relativistic effect. The direct relativistic effect significantly decreases the energy of the $6s$ orbital in mercury (Fig. 17.1), which increases the energy gap between the $6s$ orbital and higher lying

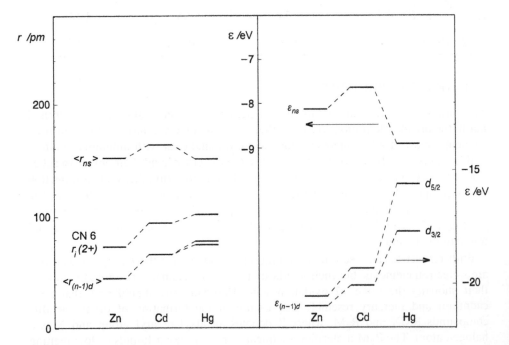

Fig. 17.1 Radii and orbital energies of Group 12 elements.

$6p$ orbitals and makes, thereby, the overlap of the s and p bands less effective and metallic bonding weaker. One can also argue that because of the direct relativistic effect in the $6s$ orbital of mercury, the I_1 and I_2 ionization potentials are high, which makes the atomization energy of mercury low, see Section 9.4. Since mercury is a liquid at room temperature it has higher electrical resistivity than its solid congeners in the Group. However, the difference between resistivities (94×10^{-8} Ω m for Hg and 6×10^{-8} Ω m for Zn) is considerably higher than that commonly observed for a metal in liquid and solid state. This suggests that the main reason for the high resistivity of mercury may be its unique band structure.

In contrast to the Group 2 elements, zinc, cadmium and mercury exhibit a strong tendency to form complexes. Since Zn^{2+} and Mg^{2+} have almost the same 6-coordinate radius and have the same formal charge, the difference with respect to complex formation between these two ions (Table 17.2) and, in general, between

Table 17.2 – Comparisons between stability constants $(\log_{10} K_1)^a$ for formation of Zn^{2+} and Mg^{2+} complexes.

Ligand		Zn^{2+}	Mg^{2+}
GROUP 17	fluoride	1.1	1.8
	chloride	0.5	(−1.0)
GROUP 16	oxalate	4.9	3.4
	acetylacetonate	5.1	3.7
	thiosulphate	2.3	1.8
	thiocyanate	1.2	−0.9
GROUP 15	ammonia	2.2	0.2
	2,2'-bipyridyl	5.1	0.5
	1,10-phenanthroline	6.4	1.2
GROUP 15/16	glycine	5.0	2.2
	nta	10.7	6.5
	edta	16.4	9.1

a Comparisons should be made with some caution, because the accuracies of some of these $\log_{10} K$ values may be no better than ± 1.

Group 12 and 2 cations is usually ascribed to the presence of a filled "soft" d shell in Group 12 ions, contrary to the filled "hard" p shell in Group 2 ions. Indeed, the hardness calculated from eq. 4.11 is only 10.9 for Zn^{2+}, while that of Mg^{2+} is as high as 32.5 eV. Lower hardness or, better, greater softness promotes covalent contributions to bonding and makes the metal−ligand bond stronger. The Zn^{2+}/Mg^{2+} difference is markedly greater for Group 15 donor ligands than for Group 16 donors, while exceptionally MgF^+ with its pairing of very hard ions is a little more stable than ZnF^+. Another factor is that discussed in Section 16.1, where the difference in complex formation between M^+ ions of Group 11 and Group 1 has been explained. According to this explanation, for the same formal charge of the cation the effective charge felt by the ligand electrons is significantly greater for

Group 12 than for Group 2 cations in the same Period. The reason is the less efficient screening of outer electrons from the nuclear charge by the filled d than by the filled p subshell. That screening is less effective for zinc than for magnesium is shown by the higher electron affinity of Zn^{2+} (18 eV) than Mg^{2+} (15 eV). Zinc, cadmium and mercury form more or less ionic compounds with the elements of Groups 16 and 17. In almost all compounds with Group 16 elements the cation shows tetrahedral coordination. They all, particularly the suphides, are insoluble in water. Among the dihalides only the fluorides show predominantly ionic character, have high lattice energies and are, therefore, only sparingly soluble in water. In contrast, the other halides of zinc and cadmium are hygroscopic and very soluble in water. Since the Zn^{2+} and Cd^{2+} ions are soft, their compounds with soft Cl^-, Br^- and I^- ligands show significant covalent character which is revealed by their relatively low melting points, and especially by the crystal structures of $CdCl_2$ and CdI_2 (see Section 13.3). The stability constant pattern for the halide complexes of Zn^{2+}, Cd^{2+}, and Hg^{2+} (Table 17.3) illustrates well the operation of the HSAB principle, with Zn^{2+} just on the hard side, Cd^{2+} just on the soft side, of the hard/soft border, and Hg^{2+} definitely soft (η = 10.9, 10.3 and 7.7 for Zn^{2+}, Cd^{2+} and Hg^{2+} respectively). A similar pattern for the Group 13 cations Al^{3+} to In^{3+} appears in Table 9.2.

Table 17.3 – Stability constants $(\log_{10} K_1)^a$ for complex formation between Zn^{2+}, Cd^{2+} and Hg^{2+} and halide ligands

	Zn^{2+}	Cd^{2+}	Hg^{2+}
fluoride	1.1	0.5	1.6
chloride	0.5	1.8	6.7
bromide	−0.6	1.9	9.4
iodide	−1.5	2.0	12.9

a Comparisons should be made with some caution, because the accuracies of some of these $\log_{10} K$ values may be no better than ±1.

All three cations give simple aqua-cations in (sufficiently acidic) aqueous solution, all three cations having primary hydration numbers of 6. Formation of series of halide complexes must therefore involve changes in cooordination number and stereochemistry, e.g. from octahedral $[Zn(H_2O)_6]^{2+}$ to tetrahedral $ZnCl_4^{2-}$ (which is indeed tetrahedral in solution, unlike $FeCl_4^-$, which is actually $[FeCl_4(H_2O)_2]^-$). Such changes in coordination number and stereochemistry are reflected in the dependence of $\log_{10} K_n$ on the number of ligands, n. Discontinuities in $\log_{10} K_n$ trends in Table 17.4 indicate where such changes occur. Thus the octahedral → tetrahedral change in the Cd^{2+}/I^- system occurs on addition of the third iodide, with an increase in stability on going from $[Cd(H_2O)_4I_2]$ to $[Cd(H_2O)I_3]^-$. The sequence for Hg^{2+}/Cl^- highlights the stability of linear $HgCl_2$ (or of tetragonal $[Hg(H_2O)_4Cl_2]$ with two short Hg–Cl and four long Hg⋯O distances) and the much lower stability of the $HgCl_3^-$ moiety.

Table 17.4 – Stepwise stability constants, $\log_{10} K_n$, for
halide complexes of Zn^{2+}, Cd^{2+}, and Hg^{2+}.

	Cd^{2+}/I^-	Hg^{2+}/Cl^-
$\log_{10} K_1$	1.9	6.7
$\log_{10} K_2$	0.8	6.5
$\log_{10} K_3$	1.7	0.8
$\log_{10} K_4$	1.3	1.0

In spite of having the same hydration number and a much higher ionic radius, the first hydrolysis constant of $Hg^{2+}aq$ ($pK_a = 2.5$) is much greater than that of $Cd^{2+}aq$ ($pK_a = 7.9$; pK_a of $Zn^{2+}aq = 9.5$ – all three values at zero ionic strength). The probable reason is relativistic stabilization of the $6s$ orbital in mercury. This stabilization attracts electrons from oxygen towards mercury in the Hg–O bond and thus facilitates proton loss. In this respect the Hg/Cd pair resembles the Tl/In pair (see Section 9.3.2). Solvent exchange and complex formation are fast at all three cations.

17.2 CHANGES IN PROPERTIES DOWN THE GROUP

Within the Group there is similarity between the chemistry of zinc and cadmium, while mercury differs considerably. The direct relativistic effect and the presence of a filled $4f$ shell, which both stabilize $6s$ electrons and thus make the ionization potentials I_1 and I_2 of mercury higher than those of its lighter congeners, are the reasons for the relatively noble character of metallic mercury. This is reflected in the high positive value of the reduction potential, $E^o(2+/0) = +0.853$ V. In contrast to mercury, metallic zinc is a strong reducing agent. The main oxidation state, as in Group 2, is the $+2$ oxidation state. Zinc, cadmium and mercury show also the $+1$ oxidation state in the M_2^{2+} dimeric cations (Zn and Cd in molten chlorides, Hg also in aqueous solution). In fluorosulphuric acid (HSO_3F) and in molten NaCl mercury even forms M_n^{2+} cations ($n = 3$ or 4), in which mercury shows the formal oxidation number $+2/n$. The remarkable stability of the Hg_2^{2+} ion, like that of the isoelectronic Au_2 molecule, arises from the direct relativistic effect which stabilizes the $6s$ electrons. The $+3$ oxidation state for Group 12 elements is disfavoured by the high values of the third ionization potential which increases across the d series faster than the ionization potentials I_1 and I_2, see Fig. 15.2.

Zn^{2+} and Cd^{2+} form complexes in which they usually display CN 4 (tetrahedral coordination), 5, or 6. Complexes having the higher coordination numbers are often in equilibrium with the tetrahedral form. A frequently encountered coordination environment of Hg^{2+} is a very distorted octahedron with two bonds much shorter then the other four. For example, in solid $HgBr_2$ there are two short and four long Hg–Br distances (248 and 323 pm respectively). The two short distances are those between the Hg atom and Br atoms in the $HgBr_2$ "molecule" and the four long distances are those between the Hg atom in $HgBr_2$ and four bromine atoms from the four other $HgBr_2$ "molecules". In many cases the distortion of the octahedron results in a genuine linear 2-coordination as e.g. in solid $Hg(CN)_2$. This consists of weakly interacting $Hg(CN)_2$ molecules and shows highly covalent bonding. The low coordination number of two in many Hg(II) compounds is probably due to the large energy gap between the $6s$ and $6p$ orbitals, which makes sp hybridization energetically more favourable than sp^2, sp^3 or sp^3d^2 hybridization. The large difference between the $6s$ and $6p$ orbital energies is the result of relativistic stabilization of the $6s$ orbital. Thus it turns out that the low coordination number as well as the low melting and boiling points of mercury may have the same origin. Easy formation of complexes and the intermediate radius of the Zn^{2+} cation make zinc one of the most important metals biologically – more than 300 enzymes containing zinc have been identified. On the other hand, cadmium and mercury are toxic elements because they form strong bonds with sulphur atoms in sulphur-containing aminoacids.

A unique property of mercury is the formation of alloys (solid or liquid under normal conditions) with several metals of Groups 1, 2, and 11, with p block metals, particularly In and Tl, and with its own lighter congeners Zn and Cd. Also the NH_4 radical which is formed by reduction of NH_4^+ dissolves in the mercury used as the cathode. On the other hand mercury does not form amalgams with most d elements, so that iron is commonly used for containers for mercury. The general rule seems to be that mercury, which has the very low atomization enthalpy of 61.3 kJ mol^{-1}, forms amalgams with metallic elements which also have low atomization enthalpies, hence not with d elements. The probable reason is that the energy released in formation of two M–Hg bonds compensates more or less for the energy loss on breaking the M–M and Hg–Hg bonds only when the atomization energies are similar. The ease of formation of an amalgam with gold, which is a basis for industrial extraction of gold from gold-bearing deposits, contradicts this rule, because ΔH_{atom} for Au is as high as 366 kJ mol^{-1}. The probable reason for amalgamation in this case is that gold, like the alkali metals, has only one s electron in the valence shell. In mercury with its two electrons per atom the s band is filled, whereas in a hypothetical 1:1 amalgam formed by an s^1 element only 3/4 of the s band would be filled. This would make possible formation of relatively strong metallic bonding, in spite of poor overlap of s and p bands in mercury.

18

Lanthanides and actinides

18.1 GENERAL CHARACTERISTICS

Because of the very small radial probability density in the vicinity of the nucleus the $4f$ orbitals drop sufficiently in energy to be filled by electrons only at Ce ($Z = 58$) and the $5f$ orbitals only at Pa ($Z = 91$), see Section 2.2. The energy of nf orbitals is particularly close to that of $(n+1)d$ orbitals at the beginning of each series, hence Ce, Pa, U and Np have mixed $4f^1 5d^1 6s^2$, $5f^2 6d^1 7s^2$, $5f^3 6d^1 7s^2$ and $5f^4 6d^1 7s^2$ configurations, respectively.* Since also within each series the $(n+1)d$ and nf orbitals do not differ much in their energies, Gd and Cm have the $f^7 d^1 s^2$ instead of the $f^8 s^2$ configuration. The tendency to accommodate the eighth electron in the $(n+1)d$ rather than in the nf shell is explained below, in Section 18.5.

By analogy with d-electron elements we should call f-electron or inner-transition elements only those elements that have partly filled $4f$ or $5f$ shells, at least in one of their common oxidation states. According to this definition lutetium and lawrencium are not f-electron elements since they have filled f shells even in oxidation state +3. On the other hand ytterbium and nobelium are f-electron elements, because in the +3 oxidation state their f shells are not filled. Since La and Lu show similar properties to elements with Z from 58 to 70, lanthanum and the fourteen elements which follow it are commonly called lanthanides. Since scandium and, especially, yttrium closely resemble lanthanides (yttrium has the same ionic radius as holmium), all these elements are sometimes grouped together under the general title "rare earths". Although the elements from thorium to plutonium differ significantly from those that follow them, particularly with respect to stability of oxidation states, all the elements from actinium to lawrencium are considered to belong to the same actinide series. The main reason is that they all show similar properties when in oxidation state +3 or +4.

Lanthanides and actinides are highly electropositive elements (χ about 1.2), hence their chemistry is predominantly ionic. Their electropositive nature means that, at least in oxidation states +2 and +3, they form salts. These are often hydrated, sometimes heavily; the solving of the crystal structure of $Nd(BrO_3)_3.9H_2O$ many years ago provided the first indication of the possibility of hydration numbers of nine. The oxocations AnO_2^+ and AnO_2^{2+} of high oxidation state (+5, +6) actinides (U, Np, Pu, Am) also form salts, in behaviour reminiscent of vanadium, compare, e.g. UO_2SO_4 with $VOSO_4$. The lanthanides and actinides all dissolve in non-oxidizing acids and react even with water. Europium and ytterbium, like Group 2 elements, dissolve in liquid ammonia to form solvated $M(NH_3)_n^{2+}$ cations. As

---

* In this Chapter the general notation for f orbitals is nf, hence the next more distant orbitals are labelled $(n+1)s$, $(n+1)p$, $(n+1)d$ and $(n+2)s$.

electropositive elements lanthanides and actinides eagerly enter into combination with electronegative p block elements, such as oxygen, nitrogen and Group 17 elements, with the formation of ionic compounds. The oxides M_2O_3 are strongly basic and the lighter ones resemble Group 2 oxides in this respect. Lanthanides and actinides also react with elements of intermediate electronegativity such as hydrogen, boron, carbon, sulphur and selenium. The products of such reactions have structures and properties intermediate between those of salts and metals. With hydrogen lanthanides form in the first stage the highly reactive and electricity-conducting solids LnH_2. These are composed of Ln^{3+}, two H^- ions, and an electron located in the conduction band; monochalcogenides, LnZ (Z = S, Se, Te), similarly consist of Ln^{3+} and Z^{2-} ions with one electron from each cation in the conduction band. These and other binary compounds are discussed further in Section 18.6.

Electronegativity, I_1, I_2 and even I_3 values of lanthanides (see Table 9.4) and actinides do not change very much across each series, which is one of the reasons why all lanthanides and heavy actinides are chemically very similar. However, I_3 and I_4 show significant maxima and minima in the middle of each series (see Figs. 18.7 and 18.8 below), which affect some chemical properties of the relevant elements. On the other hand light actinides from Th to Pu differ considerably from the rest of the $5f$ elements, and Th, Pa and U even show some resemblance to Group 4–6 elements, respectively.

18.2 THE METALLIC PHASE

Light lanthanide and actinide metals abound in allotropic modifications, but the end members of each series only occasionally show more than one modification. This behaviour is commonly attributed to participation of f orbitals in metallic bonding at the beginning of each series, particularly in the early actinides. Melting points of lanthanides and actinides are rather low and only in a few cases exceed 1500 °C. In this respect f elements differ considerably from d elements, particularly from those in the second and third transition series. Beginning with cerium, melting points of lanthanides increase across the series, except for europium and ytterbium, which show much lower melting points than their neighbours. The reason for this unique behaviour is that europium and ytterbium, which have the $4f^7 6s^2$ and $4f^{14} 6s^2$ configuration, donate only two electrons to the energy band, in contrast to the rest of lanthanides which donate three electrons. A smaller number of electrons in the band means weaker metallic bonding, hence lower atomization enthalpy, and lower melting and boiling points. Because of their low ΔH_{atom}, Eu and Yb, in contrast to the rest of the lanthanides, easily form amalgams (see Section 17.2). Contribution of only two electrons to the band by Eu and Yb is a manifestation of the high value of I_3. High I_3 hinders the transfer of the f electron to the energy band where it becomes delocalized. The unique properties of the f^7 and f^{14} configurations with respect to the relevant ionization potentials have the same origin as the unique properties of the d^5 and d^{10} configurations, discussed in Sections 15.6 and 16.1. The metallic radius of lanthanides decreases across the series more or less uniformly, again with the exception of Eu and Yb which show very high metallic radii. Donation of only two electrons to the band leaves the atomic cores larger, hence the greater radius and lower binding forces.

Lanthanides and actinides, like transition elements, form *s/p* and *d* energy bands. In addition, the *f* orbitals form a low-lying and narrow band. This band is narrow because *f* orbitals have very small radial extent (Figs. 18.1 and 18.2), which restricts their overlap on adjacent atoms. Strong evidence for participation of 5*f* orbitals in metallic bonding in actinide metals is the non-monotonic change in the metallic radius across the actinide series, Fig. 18.3. The significant decrease in the metallic

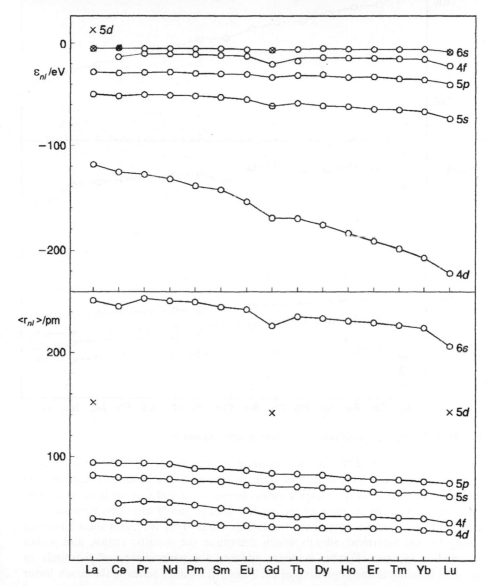

Fig. 18.1 Energy and spatial order of orbitals in lanthanide atoms.

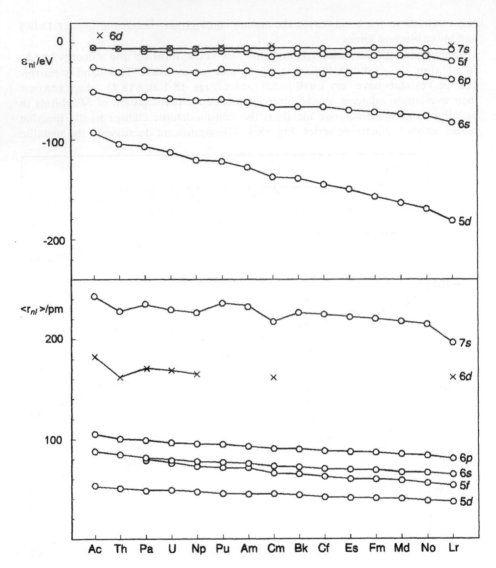

Fig. 18.2 Energy and spatial order of orbitals in actinide atoms.

radius between Ac and Np can be explained by poor shielding from the increasing nuclear charge by delocalized $6d$ and $5f$ electrons. In order to explain the unexpected increase in the metallic radius between Np and Cm we have to assume that with continuing increase of the nuclear charge the $5f$ electrons become more and more localized i.e. they leave the energy band. This increases their shielding ability for the outermost orbitals which determine the metallic radius, and makes this radius increase. Further evidence adduced for participation of f orbitals in metallic bonding at the beginning of each inner-transition series is the much lower melting temperatures of lanthanum and actinium, in comparison with those of

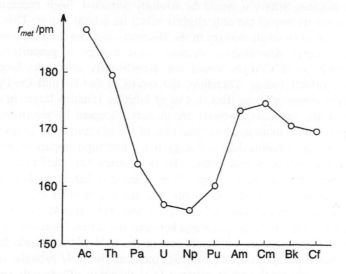

Fig. 18.3 Metallic radii of actinides.

scandium and yttrium. That the magnetic properties of U, Np and Pu differ from those of Cm and Bk also points to a contribution from f orbitals to metallic bonding in the light actinides.

18.3 THE SEQUENCE OF ORBITAL ENERGIES AND RADII

An important property of $4f$ orbitals is that with respect to radius they are located below the $5s$ and $5p$ orbitals, whereas their energy is less negative. In other words, the $4f$ orbitals are spatially below the $5s$ and $5p$ orbitals, but are higher with respect to energy, Fig. 18.1. It can be seen in Fig. 18.2 that the $5f$ orbitals behave similarly. This singular behaviour is commonly explained by changes in orbital energy with Z – different for f than for s and p orbitals. In light atoms the energy of vacant nf orbitals is more negative than that of vacant $(n+1)s$ and $(n+1)p$ orbitals. This is because the outermost orbitals behave like hydrogen orbitals with their energy depending almost solely on the principal quantum number n. However, we know from Chapter 1 that the radial probability density of finding an electron in the vicinity of the nucleus depends on the quantum number l in the following way :

$$P_{nl}(r) \propto (r/a_0)^{2l} r^2 .$$

From this expression $P_{nl}(r)$ near the nucleus is much greater for the s and p than for the f electron with the same n quantum number. This makes the energy of s and p orbitals decrease very rapidly with increasing nuclear charge, whereas the energy of f orbitals remains almost constant. Therefore, at relatively low atomic numbers the energy of the vacant $(n+1)s$ and $(n+1)p$ orbitals drops below that of the vacant nf orbitals in spite of the higher principal quantum number. For $4f$ orbitals this happens at Z equal to about 25. In a more pictorial way one can say that an electron when placed in the $(n+1)s$ or $(n+1)p$ orbital would make frequent "excursions" toward the nucleus, where it would be strongly attracted. Such excursions would greatly decrease its energy but only slightly affect its orbital radius. This is because near the nucleus even small changes in the electron – nucleus distance would change the electron energy dramatically, because there energy is proportional to $1/r$. However, such small changes would not significantly affect the large average distance, i.e. orbital radius. Therefore, the radius of $(n+1)s$ and $(n+1)p$ orbitals, which in light atoms is larger than that of nf orbitals, remains larger in all heavier atoms (where the respective orbitals are already occupied by electrons), whereas their energy becomes more negative than that of the nf orbitals. A different order of orbitals in respect to radius than to energy has a profound impact on properties of lanthanides and actinides and makes the two series very different from the d elements. The filling of inner f orbitals by electrons leaves in lanthanides and actinides, in contrast to d block elements, the number of outer electrons constant, which is the main reason why the M^{3+} (and also M^{2+} and M^{4+}) lanthanide and actinide cations differ so little across and between the series. Moreover, the outer $5s$ and $5p$ ($6s$ and $6p$) electrons efficiently screen the field of the ligands felt by the f electrons. Therefore, splitting of f orbitals, particularly of $4f$ orbitals, in a ligand environment is very weak and, in contrast to d elements, affects only very slightly the chemical and spectroscopic properties of f element cations. For instance, when electronic transitions from one J state to another J state, called f-f transitions, occur the absorption bands are very sharp and in general show little dependence on the ligand, contrary to intraconfigurational d-d transitions. Nevertheless, the effect of a crystal field cannot be always ignored. The intensities of a number of bands called hypersensitive bands show a distinct dependence on the coordinated ligand.

Lanthanide ions, except La^{3+} and Lu^{3+}, show magnetic moments which, in contrast to d block elements, are not spin-only but contain a contribution from the orbital angular momentum. The reason is that the ligand field acting on the $4f$ orbitals, which are shielded from the outer field by $5s$ and $5p$ orbitals, is too weak to quench the orbital angular momentum. Such quenching happens with d orbitals which are outer orbitals for the ions and are, therefore, subject to strong action of the electrostatic field of the ligands. The magnetic properties of actinides, particularly of light actinides, are intermediate between those of lanthanides and of d electron elements. This is because the spatial extent of $5f$ orbitals is significantly greater than of $4f$ orbitals and they are, therefore, less efficiently shielded by the outer $6s$ and $6p$ orbitals from the ligands' electrostatic field. Moreover, at the beginning of the actinide series $5f$ orbitals participate in bonding, affecting in this way the magnetic properties of early members.

In some f-electron atoms not only the f orbitals but also d orbitals show unusual behaviour in respect to radius and energy. In the Lu, Ac–Np, Cm and Lr atoms the $(n+1)d_{3/2}$ orbitals, in spite of their less negative energy, have a smaller radius than the $(n+2)s$ orbitals, Fig. 18.4. The different sequence of radii and energies of s and d orbitals is found in all actinides which have $6d$ electrons. Instead, among lanthanides which have $5d$ electrons only lutetium shows such a reversal of radii and energies. This difference points to the importance of relativistic effects which, because of the greater atomic number, influence the valence s and d electrons more strongly in actinides than in lanthanides. Indeed, calculations show that for nonrelativistic orbitals the less negative orbital energy corresponds with larger radius and *vice versa*. Since stabilization of *s* electrons and destabilization of *d* electrons is particularly strong in heavy atoms, different ordering of $6d$ and $7s$ orbitals with respect to energy and radius is a common feature of transactinide elements (see Chapter 19).

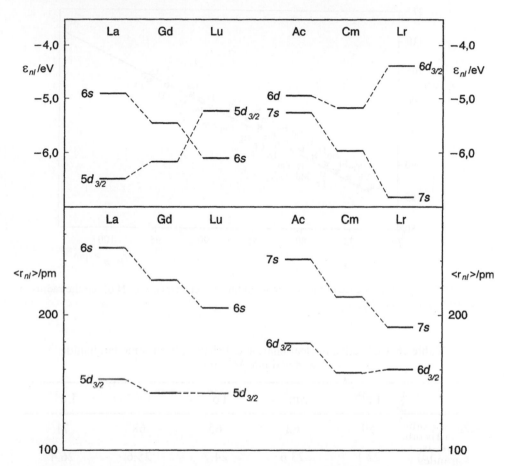

Fig. 18.4 Energy and spatial order of outer $d_{3/2}$ and *s* orbitals in atoms of some lanthanides and actinides.

18.4 LANTHANIDE AND ACTINIDE CONTRACTIONS

It can be seen in Figs. 18.1 and 18.2 that with increasing atomic number the radii of nd, nf and all more outlying orbitals in lanthanides and actinides contract. As we know, orbitals contract across each row of the Periodic Table, because of incomplete shielding from nuclear attraction by inner electrons and by electrons in the same subshell. Of particular importance to the chemistry of the lanthanides and actinides is the decrease of the radius of the $(n+1)p$ subshell, because this subshell determines the radius of the M^{3+} cations. This follows from Fig. 18.5, which shows that the ionic radius decreases linearly with the radius of the $(n+1)p$ subshell and exceeds the latter by only ~ 10 pm. Contraction of the $(n+1)p$ subshell in ions, hence also of the ionic radius of lanthanides and actinides, is not particularly large. This is because the effective nuclear charge acting on the $5p$ and $6p$ subshells does not

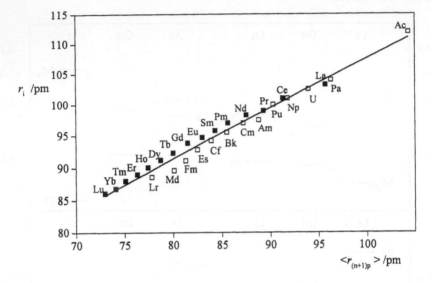

Fig. 18.5 The dependence of the M^{3+} ionic radius of f elements, r_i (CN 6), on the radius of the $(n+1)p$ subshell, $\langle r_{(n+1)p} \rangle$

Table 18.1 Effective nuclear charge and shell radii in some lanthanide and actinide M^{3+} ions.

	Pr^{3+}	Sm^{3+}	Tb^{3+}	Er^{3+}	Lu^{3+}
$+Ze$	59	62	65	68	71
$+Z_{eff}e\,(5p)$	22.1	23.6	24.7	25.6	26.4
$\langle r_{5p} \rangle$ / pm	89.3	84.3	80.0	76.3	73.0
	Pa^{3+}	Pu^{3+}	Bk^{3+}	Fm^{3+}	Lr^{3+}
$+Ze$	91	94	97	100	103
$+Z_{eff}e\,(6p)$	29.1	31.0	32.8	34.5	36.0
$\langle r_{6p} \rangle$ / pm	96.3	90.3	85.5	81.3	77.8

change much across each series, Table 18.1. One can see in this Table that between Pr^{3+} and Lu^{3+} the nuclear charge increases by 12 units, whereas the effective nuclear charge for $5p$ orbitals increases by only 4.3 units and the $5p$ radius decreases by 16.3 pm. For the same increase in the atomic number of actinides the effective nuclear charge for $6p$ orbitals increases by 6.9 units and the $6p$ orbital radius decreases by 18.5 pm. Because the radius of the $6p$ subshell decreases somewhat faster than that of the $5p$ subshell, the ionic radii of lanthanides and actinides converge across the series (see Fig. 18.6). The probable reason why $Z_{eff}(6p)$ and the $\langle r_{6p} \rangle$ radius change faster than the respective values for lanthanides is the significantly greater radial extent of $5f$ than $4f$ orbitals. Greater radial extent means less efficient shielding of the more distant shells from the nuclear charge.

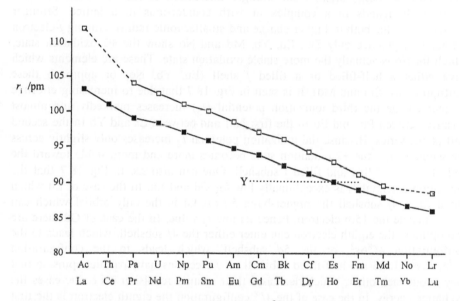

Fig. 18.6 Radii of lanthanide (■) and actinide (□) M^{3+} ions.

The decrease in the $r_i(3+)$ radii, called the lanthanide or actinide contraction, is moderate and for one added electron is much smaller than contraction of M^{3+} cations of transition elements. For instance, upon addition of one f electron the $r_i(3+)$ radius of lanthanides decreases on the average by only 1.15 pm. On the other hand the average decrease in the $r_i(3+)$ radius of d block elements on addition of one d electron (only for ions with empty e_g orbitals) is equal to about 3.5 pm (see Fig. 15. 10). Relatively small contraction, constant number of electrons in the outer-most $(n+1)p$ shell in the ion, and a very small ligand field effect are the main reasons why lanthanide and heavy actinide M^{3+} ions are so similar to each other, and why their separation is so difficult. However, the small decrease in the ionic radius is large enough for geochemical processes to separate lanthanides with respect to their occurrence into a fraction that contains essentially only light lanthanides and a second fraction which contains mainly yttrium and heavy lanthanides.

18.5 OXIDATION STATES OF LANTHANIDES AND ACTINIDES

In spite of having two electrons in the outermost valence shell (the $6s$ or $7s$ subshell) lanthanides and heavy actinides, in contrast to Group 2 elements, as a rule show a stable +3 oxidation state. The reason is that for most f elements the third electron is detached from the nf orbital, which has orbital energy not very different from that of the $(n+2)s$ orbital. On the other hand, in the case of Group 2 elements the third electron has to be removed from the very strongly bonded $(n-1)p$ shell. With respect to factors stabilizing the +3 oxidation state lanthanides and actinides resemble scandium and yttrium, which also have two s electrons in the outermost shell and show an oxidation number of +3, see Section 15.3. However, additional energy is necessary to attain the +3 oxidation state in lanthanides and actinides. This compensating energy arises from stronger interaction of the M^{3+} than the M^{2+} cation with ligands in a complex or with counter-ions in a lattice. Stronger interaction is due both to higher charge and smaller ionic radius. Among f-electron elements in practice only Sm, Eu, Yb, Md and No show the +2 oxidation state, which for No is actually the more stable oxidation state. These are elements which have either a half-filled or a filled f shell (Eu, Yb, No) or approach these configurations (Sm and Md). It is seen in Fig. 18.7 that due to increasing effective nuclear charge the third ionization potential (I_3) increases markedly and almost linearly between Pm and Eu in the first half and between Er and Yb in the second half of the series. Because the ionization potential I_2 increases only slightly across the whole series, the +2 oxidation state becomes more and more stable toward the end of the half-filled and filled f subshell. One can also see in Fig. 18.7 that the ionization potential I_3 is exceptionally low for Gd and Lu. In the case of Lu which has a filled $4f$ subshell the higher-lying $5d$ orbital is the only orbital which can accommodate the 15th electron, hence its low I_3 value. In the case of Gd there are two options: the eighth electron can enter either the $4f$ subshell, which leads to the configuration $4f^8 6s^2$, or the $5d$ subshell, which leads to the configuration $4f^7 5d^1 6s^2$. Up to the half-filled subshell all f electrons have parallel spins so that they avoid each other, which decreases their mutual repulsion – i.e. increases the exchange energy. In the case of the $4f^8$ configuration the eighth electron is the first to have its spin opposed to those which already populate the $4f$ subshell. This allows the eighth electron to appear in close vicinity to the other seven and experience additional interelectronic repulsion, which destabilizes the system (see also Section 15.6). To avoid the additional repulsion the Gd atom adopts the $4f^7 5d^1 6s^2$ configuration. However, since the energy gap between $5d$ and $4f$ orbitals is not large this $4f^7 5d^1 6s^2$ configuration is not much more stable than the $4f^8 6s^2$ configuration. Therefore, the ionization potential I_3 is almost as low as if the eighth electron were detached from the $4f$ subshell and a sudden drop in I_3 is observed between Eu and Gd, similar to that between Mn and Fe or Re and Os (again see Section 15.6). A similar (but smaller) drop, for the same reason, is observed in the actinide series between Am and Cm. In contrast to europium, its homologue americium does not show the oxidation number +2. The reason is that although both elements in the +2 oxidation state have the same $f_{5/2}^6 f_{7/2}^1$ configuration it is easier to detach an electron from the less strongly bonded $5f$ than from the $4f$ subshell. Moreover the

$5f_{7/2}$ electron in Am(II) is relativistically more destabilized than the $4f_{7/2}$ electron in Eu(II), which should additionally decrease the third ionization potential of americium. Indeed, the third ionization potential of Am is only 0.6 eV higher than that of Pu, whereas the respective difference for the Eu–Sm pair is 1.5 eV.

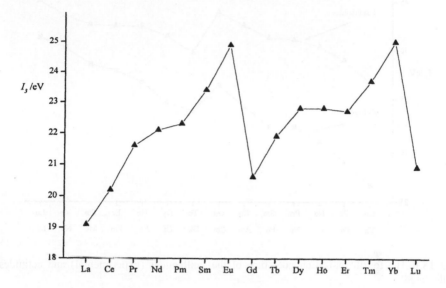

Fig. 18.7 The third ionization potentials of the lanthanides.

Lanthanides as a rule do not show the oxidation number +4 (or higher), because they have high fourth ionization potentials (I_4), see Fig. 18.8. Only the first members Ce and Pr, and also Tb, have relatively low I_4 values and display, therefore, the oxidation number +4. One can see that, except for the last member, plots in Fig. 18.8 have a similar shape to that in Fig. 18.7, but are shifted by one element to the right. Therefore, we now observe a maximum at Gd instead at Eu and a minimum at Tb instead at Gd. In the case of the actinides the maximum is shown by Cm and the minimum by Bk. However, the number of f electrons in the ion from which the next electron is detached is the same as in the plot in Fig. 18.7, i.e. 7 at the maximum and 8 at the minimum (the latter if we assume that the configuration of Gd^{2+} is $4f^8$). Therefore, the sudden drop in I_4 between Gd and Tb and between Cm and Bk can be explained in the same way as the drop in I_3 between Eu and Gd and between Am and Cm. In contrast to lanthanides, light actinides, which all have low values of the fourth ionization potential, I_4, readily form the +4 oxidation state. Because the value of I_4 is low for terbium and berkelium these two elements (in particular berkelium) show the +4 oxidation state, mainly in oxides and fluorides. Also in contrast to light lanthanides, light actinides, depending on the number of d and f valence electrons, additionally display oxidation numbers from +5 for Pa to +7 for Np and Pu. In the oxidation states +5 and +6 U, Np, Pu and Am form linear dioxo ions AnO_2^+ and AnO_2^{2+}, respectively. The main reason for the difference between light actinides and light lanthanides with respect to I_4 and higher

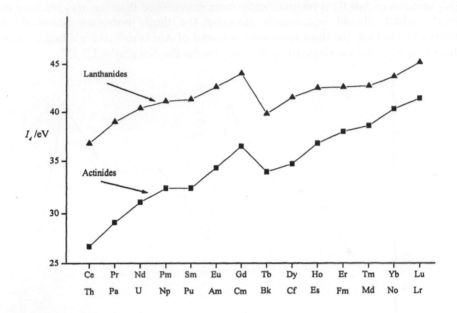

Fig. 18.8 The fourth ionization potentials of the lanthanides (experimental) and actinides (calculated).

ionization potentials is probably repulsion between the first and next orbital with the same l quantum number (see Sections 3.2 and 15.6). In this case the repulsion is between the $4f$ and $5f$ orbitals. A measure of this repulsion is the distance between the $nf_{5/2}$ and nd orbitals, which is much greater for e.g. plutonium than for its homologue samarium, Fig. 18.9 (compare with Figs. 3.3 and 15.15).

The much greater radius of $5f$ than of $4f$ orbitals is paralleled by less negative $5f$ orbital energy and by lower ionization potentials involving $5f$ electrons. However, calculations show that the effective nuclear charge, $+Z_{eff}e$, acting on f electrons increases much faster across the actinide than the lanthanide series. For instance, $+Z_{eff}e(5f)$ increases from 21.6 to 32.3 in the interval between Pa^{3+} to Lr^{3+}, whereas $+Z_{eff}e(4f)$ increases from 18.4 for Pr^{3+} to only 25.8 for Lu^{3+}. The probable reason is the greater radial extent of $5f$ than $4f$ orbitals, which may bring about less effective mutual shielding of $5f$ than $4f$ electrons from nuclear attraction. The rapidly increasing effective nuclear charge acting on $5f$ electrons is the reason why the heavy actinides almost catch up with the lanthanides with respect to the fourth ionization potential (see Fig. 18.8) and, like the lanthanides, have a stable +3 oxidation state (except for No).

18.6 BINARY COMPOUNDS AND SALTS

Both lanthanides and actinides form numerous binary compounds and salts in almost all their various oxidation states. The lanthanides and right-hand actinides favour oxidation state +3, though there are significant numbers of M(II) and M(IV),

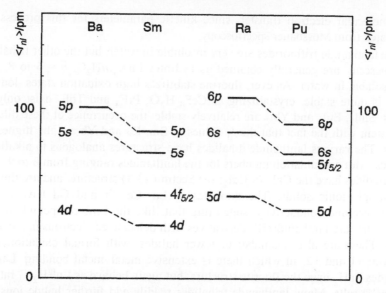

Fig. 18.9 The difference in the radial extent between $nf_{5/2}$ orbitals and nd subshells in Sm and Pu. The radii of closed subshells are average values weighted over the occupancy of $p_{1/2}$, $p_{3/2}$ and $d_{3/2}$, $d_{5/2}$ orbitals.

and indeed of mixed valence, compounds. Some of the earlier actinides resemble d block elements in the ranges of oxidation states they exhibit. Thus Th, Pa, U and Np have highest oxidation states of +4, +5, +6 and +7 respectively, paralleling Ti, V, Cr and Mn and their second and third row analogues.

The lanthanide sesquioxides Ln_2O_3 are all well known. They occur in a variety of structures, with the lanthanide having CN 6 or 7. Dioxides are known for Ce, Pr, and Tb; at least four monoxides exist. The dioxides and monoxides have the fluorite and sodium chloride structures respectively. There are also a number of mixed valence oxides, such as Pr_6O_{11} and Tb_4O_7. Sesquioxides are again important for the actinides, at least from Pu_2O_3 onwards – higher oxides are more important for the earlier actinides. Indeed for Np and Pu it is oxygen which brings out the highest oxidation state, of +7 in the mixed oxides Li_3AnO_5. ThO_2 has a particularly high melting point, 3 663 K; binary borides, carbides, silicides and nitrides of the actinides are also very refractory materials. These binary compounds exhibit a range of stoichiometries; the 1 : 1 compounds AnC and AnN have the NaCl structure and show some salt-like properties.

Most of the lanthanides form compounds Ln_2Z_3 and LnZ with Z = S, Se, Te. Ln_2S_3, and An_2S_3, have a variety of structures, with Ln or An having CN 6 (for smallest Ln, viz. Lu, Yb), 7 (in Gd_2S_3) or 8 (e.g. La_2S_3, Ac_2S_3, Pu_2S_3). The LnZ compounds have the sodium chloride structure, but are not simple ionic compounds. Most are black solids with metallic conductivity corresponding to the formulation $Ln^{III}Ze^-$ (cf. p.170). The sulphides Ln_3S_4 have the 8:6-coordinated Th_3P_4 structure, but whereas La_3S_4 and Ce_3S_4 behave as $Ln^{III}_3S_4e^-$, Sm_3S_4 and Eu_3S_4 are mixed valence compounds $Ln^{II}Ln^{III}_2S_4$. Eu_3S_4 is of considerable interest in relation to intramolecular electron transfer, since kinetic parameters for this process can be estimated from Mössbauer spectroscopy.

The lanthanide trifluorides are very insoluble in water, but the other trihalides are deliquescent, are generally obtained as hydrates $LnX_3.nH_2O$, $n = 6$ to 8, and are very soluble in water. As ever, fluorine stabilizes high oxidation states, iodine low. CeF_4 is quite stable, crystallizing as $CeF_4.H_2O$; PrF_4 and TbF_4 exist only as dry solids. SmI_2, EuI_2 and YbI_2 are relatively stable; the occurrence of these dihalides is consistent with the fact that the I_3 values of Sm, Eu and Yb are the highest in the series. The various lanthanide dihalides have structures analogous to alkaline earth halides, with coordination numbers for the lanthanides ranging from 6 to 9. Some of the diiodides have the CaI_2 ($\equiv CdI_2$, see Section 13.3) structure, and are thus clearly not simple ionic solids. The diiodides of La, Ce, Pr and Gd have a metallic appearance and conductivity, suggesting that, like the LnZ compounds mentioned above, they are lanthanide(III) derivatives with delocalized electrons in a conduction band. There are also a number of lower halides, with formal oxidation numbers between +1 and +2, in which there is extensive metal–metal bonding. Lanthanide hydrides LnH_2 have the fluorite structure, but again behave as $Ln^{III}H_2e^-$, rather than as Ln(II) salts. Many lanthanide trihalides readily add further halide ions to form complexes, e.g. $LnCl_6^{3-}$ from $LnCl_3$. Analogous behaviour in the actinide series leads to such series as that of the An(IV)-fluoride sequences $[AnF_5]^-$, $[AnF_6]^{2-}$, $[AnF_7]^{3-}$, $[AnF_8]^{4-}$, established for Th through to Bk. Both here and in actinide tri- and tetra-halides coordination numbers of 8 (most $AnBr_3$, AnI_3; all AnF_4) and 9 (most AnF_3, $AnCl_3$) are normal. The relative stability of the americium dihalides reflects the f^7 configuration of Am(II); thorium diiodide resembles the lanthanum diiodides mentioned above in having the thorium in its most stable oxidation state – it is effectively $Th^{IV}I_2e^-_2$. In relation to uranium, it is interesting to note formal analogies in halide chemistry to the Group 6 and Group 16 elements (cf. Chapter 12). Thus UF_6 may be compared with WF_6 and MoF_6, even with SF_6, while there are also parallels amongst oxohalides MOX_4 and MO_2X_2 between U and Cr/Mo/W and S/Se/Te.

Lanthanide and actinide cations, including AnO_2^+ and AnO_2^{2+}, form a multitude of salts with oxoanions. Many of these are hydrated, sometimes heavily. Quite often for lanthanides the highest hydrate has $9H_2O$, as for example in $LaI_3.9H_2O$, $Ce(ClO_4)_3.9H_2O$, $Nd(BrO_3)_3.9H_2O$, $La_2(SO_4)_3.9H_2O$, and the series of trifluoro-methyl- and ethyl-sulphonates $Ln(SO_3CF_3)_3.9H_2O$ and $Ln(SO_3Et)_3.9H_2O$. Usually these contain $[Ln(H_2O)_9]^{3+}$ cations, though the sulphate contains Ln^{3+} in both CN 9 and CN 12. Less heavily hydrated salts may contain $Ln^{3+}(An^{3+})$-anion-H_2O complexes, as in $GdCl_3.6H_2O$ and in $AmCl_3.6H_2O$ and its Bk analogue, where the metal ion has CN 8, in the form of $[MCl_2(H_2O)_6]^+$ (cf. e.g. $NiCl_2.4H_2O \equiv [NiCl_2(H_2O)_4]$). The colours of hydrated lanthanide cations are varied – e.g. Pr^{3+}aq is green, Nd^{3+}aq lilac, Sm^{3+}aq yellow – indicating that ligand field effects may be small but are not negligible.

The lanthanides and actinides form many double salts, such as $La(NO_3)_3.2MNO_3.4H_2O$, $Ce(SO_4)_2.2M_2SO_4.2H_2O$ and $U(ox)_2.2K_2(ox).5H_2O$ (M = an alkali metal cation; ox = oxalate). Most of these are actually complexes. Thus the ceric ammonium sulphate much used in volumetric analysis is ammonium hexanitratocerate(IV). Here, and in a thorium(IV) analogue, the nitrates are all bidentate, giving a CN of 12 for the Ce^{4+} and Th^{4+}.

18.7 AQUA-CATIONS

Tervalent lanthanides in aqueous solutions form aqua-complexes $[M(H_2O)_n]^{3+}$ with hydration numbers of 9 at the beginning of the series and 8 at the end. The tendency to hydrolysis is small for these rather large 3+ cations, increasing with increasing atomic number i.e. with decreasing ionic radius and hydration number. The Ln^{3+} ions have substantial ion hydration enthalpies, increasing steadily from Ce^{3+} (-3369 kJ mol^{-1}) to Lu^{3+} (-3759 kJ mol^{-1}) – compare these values with e.g. -4661 kJ mol^{-1} for Al^{3+}. The high coordination numbers and fairly large ionic radii however mean that the primary hydration shell water molecules are loosely held, so that water exchange and complex formation normally take place extremely rapidly. Thus rate constants for water exchange are of the order of 10^8 to 10^9 s^{-1} at ambient temperatures. Complex formation with mono- and bi-dentate ligands is comparably rapid, but with multidentate polyaminocarboxylate ligands such as dtpa, and especially with macrocyclic ligands with pendant arms or potentially encapsulating ligands, reaction rates are orders of magnitude less. Half-lives for complex formation increase from micro- or nano-seconds to minutes, hours, or, with suitable ligand tailoring, days (cf. Cu^{2+}aq, p. 158).

Consistent with the discussion in the preceding section, such 2+ and 4+ lanthanide aqua-ions as exist in aqueous solution are strongly reducing and oxidizing respectively. Reduction potentials are $E°(3+/2+) = -1.55, -0.35,$ and -1.05 V for samarium, europium, and ytterbium, $E°(4+/3+) = +1.72$ V for cerium (in acidic media in the absence of complexing anions). The high reduction potential and marked colour change at the endpoint make Ce(IV) a valuable oxidant for volumetric analysis. Ce(IV), like peroxodisulphate, is a powerful but slow oxidant; both have been used extensively in kinetic studies of oxidation of a great variety of organic substrates. Complementarily, the moderately strong reducing power of Eu^{2+}, and its ready availability by zinc amalgam reduction of Eu^{3+}, have made it a useful reductant for kinetic studies of electron transfer.

f-Block elements provide the only properly authenticated examples of 4+ aqua-cations, in the form of Ce^{4+}aq and analogues for several early actinides, such as Th^{4+}aq and U^{4+}aq. These aqua-cations exist, of course, only in strongly acidic media; pK_a values are about 1 for U^{4+}aq, 1.5 for Pu^{4+}aq, and perhaps as low as zero for Ce^{4+}aq. In weakly acidic media there will be mixtures of hydrolyzed and polymerized species. The U–O distance in U^{4+}aq in solution has been determined by X-ray diffraction as 242 pm. No estimate of a hydration number for U^{4+} appears to be available, but values of 9 and 10 have been estimated for Th^{4+}aq, and similarly high values suggested for 3+ ions of the early members of the actinide series. The dioxo-cations AnO_2^+ and AnO_2^{2+} may be deemed to be the aqua-cations of actinides in the 5+ and 6+ oxidation states. UO_2^{2+}aq is believed to have five equatorial water molecules, which undergo exchange with a rate constant of 8×10^5 s^{-1}. This exchange takes place some 10^{14} faster than exchange of the O atoms of the UO_2^{2+} unit – a dramatic illustration of the strength of An=O bonds in these moieties. In the special case of plutonium, all four aqua-cations, $Pu^{3+}aq$, $Pu^{4+}aq$, PuO_2^+aq, and $PuO_2^{2+}aq$, can coexist in aqueous solution, so similar are the various connecting reduction potentials :

$$Pu^{3+} \xleftarrow{\;+1.01\,V\;} Pu^{4+} \xleftarrow{\;+1.04\,V\;} PuO_2^+ \xleftarrow{\;+1.02\,V\;} PuO_2^{2+}.$$

$U^{3+}aq$ is a strong reducing agent ($E^o(4+/3+) = -0.52$ V). Electron transfer to reductants with potential bridging groups, e.g. $[Co(NH_3)_5X]^{2+}$ with X = e.g. Cl, NCS, or N_3, occurs by the inner-sphere mechanism.

18.8 COMPLEX FORMATION AND SEPARATION

In contrast to d-electron elements, tervalent lanthanides and actinides form only weak complexes with halogen ligands. The most stable complexes are those with chelating ligands which contain oxygen donor atoms, such as hydroxo acids and β-diketonates, and polyaminocarboxylate ligands such as edta which contain additional nitrogen donor atoms. Coordination numbers, again in contrast to d-block elements, are as a rule greater than 6. The predominant coordination numbers in lanthanide complexes are 7, 8 and 9 but complexes with CN 10 or 12 (e.g. in $[Ce(NO_3)_6]^{3-}$), or even 14 (as in solid $U(BH_4)_4$, which contains two η^3 and four bridging η^2 BH_4^- ligands around each U atom) occur for the early members of the lanthanide and actinide series. Oxoanions such as nitrate and acetate often act as bidentate ligands in complexes of f-block elements. High coordination numbers at the beginning of the lanthanide and actinide series can be attributed to both large radius of the $5p$ ($6p$) subshell which determines the ionic radius and probably also to participation of $4f$ ($5f$) orbitals in formation of hybrid orbitals.

With decreasing radius of the M^{3+} ions the stability of complexes usually increases across each series, in other words the standard free energy of complex formation, ΔG^o_{compl}, becomes more negative. Because of the small contraction in ionic radius and constant number of electrons in the outermost shell the decrease in ΔG^o_{compl}, i.e. the increase in stability constants, is rather small. For instance, even for strong complexes the average ratio of the stability constant β_3 for neighbouring lanthanides rarely exceeds two. Because the individual pairs differ considerably with respect to the ratio of stability constants, the decrease in ΔG^o_{compl} is not a smooth function of Z, Fig. 18.10. This Figure shows relative changes (with respect to lanthanum) in the average value of ΔG^o_{compl} for a dozen or so strong 1:3 lanthanide complexes with multidendate ligands. One can see in Fig. 18.10 that the f^7 configuration (Gd^{3+}) divides the lanthanide series into two subgroups: La–Gd and Gd–Lu. Within each subgroup ΔG^o_{compl} changes in a similar way. Moreover, one can also see that the f^3, f^4 configurations (Nd^{3+} and Pm^{3+}) divide the first subgroup and the f^{10}, f^{11} configurations (Ho^{3+} and Er^{3+}) the second subgroup into two segments. Within each segment ΔG^o_{compl} changes in the same way. Strong complexes of actinide M^{3+} ions also show a similar pattern. Because of the remarkable symmetry in the changes of ΔG^o_{compl} the effect has been called the double-double or tetrad effect. That the f^7 configuration is unique results from the exchange energy which in the first half of the series has its maximum just at f^7.

Calculations show that ground terms of the atoms with the f^3, f^4, f^{10} and f^{11} configurations are also stabilized, however not by coupling of spins but by coupling of orbital momenta. Because the four configurations are stabilized in both higher and lower oxidation states (either three and two or four and three) the respective ionization potentials (I_3 and I_4) show a slow increase and even decrease around the middle of the first and second half of the lanthanide and actinide series, see Figs. 18.7 and 18.8. Stabilization of the f^3, f^4, f^7, f^{10} and f^{11} configurations in ions depends on environment and decreases with increasing ability of the ligand to delocalize f electrons i.e. decreases with increasing nepheloauxetic effect. Since the ability to delocalize is very low for water (H_2O is only second to F^-) and complex formation consists in the transfer of the cation from water to ligand surroundings, lower stabilization in the ligand surroundings means a somewhat lower tendency to form complexes by those lanthanide and actinide M^{3+} ions which show one of these five configurations, Fig. 18.10.

Because of low separation factors (low ratio of respective stability constants for neighbouring elements), the separation of lanthanides is not an easy task and requires a large number of theoretical plates in column chromatography or a large number of stages in liquid-liquid solvent extraction. In ion-exchange chromatography multidentate ligands such as β-hydroxyisobutyric acid or ethylenediamine tetraacetic acid (edta) are used as eluting agents, while in the liquid-liquid extraction method tributylphosphate (tbp), di-2-ethylhexyl orthophosphoric acid (hdehp) or long-chain tertiary or quaternary alkyl ammonium salts are used. It can be seen in Fig. 18.10 that separation of the Pr–Nd, Eu–Gd, Dy–Ho and Yb–Lu pairs must be particularly difficult, because of low separation factors. Indeed even with hdehp as extractant the separation factor for the Eu–Gd pair is only about 1.5. On the other hand the separation factor in the same system

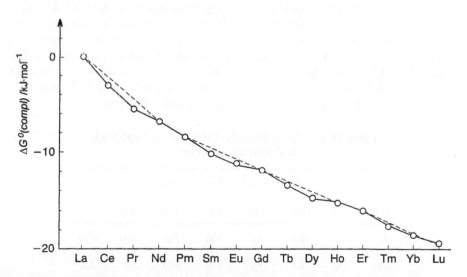

Fig. 18.10 The average standard free energy of complex formation by Ln^{3+} ions (with respect to La^{3+}) for a number of complexes with polydendate ligands.

for the Gd–Tb pair is as large as about 4. However, europium can be separated from the rest of lanthanides by reducing Eu^{3+} to Eu^{2+}, and cerium can easily be separated by making use of its relatively stable +4 oxidation state.

An important practical problem, which is also very interesting from the chemical point of view, is the separation of yttrium from the lanthanides. With respect to ionic radius the position of yttrium is equivalent to that of holmium, whereas with respect to stability constants of its complexes yttrium shows so-called nomadic properties. For most ligands, usually hard, its position is also among heavy lanthanides, whereas for some, mostly soft ligands such as the SCN^- ion, yttrium becomes a light pseudolanthanide. By applying in turn the two types of ligands yttrium can be very efficiently separated from all the lanthanides. The reason for the migration of yttrium across the lanthanide series is probably the contribution from covalence to bonding, which may result mainly from participation of f orbitals in the metal-to-ligand bonds formed by light lanthanides. Contribution from covalence makes bonding stronger in comparison to purely ionic bonding. Because yttrium has no f orbitals the contribution from covalence to bonding in complexes with soft ligands is greater for lanthanides than for yttrium. With respect to stability constants this moves light lanthanides toward yttrium (or *vice versa*), thereby making yttrium a light pseudolanthanide.

The singular role of the f^7 configuration makes it possible to arrange lanthanides and actinides with respect to some properties into a kind of "Periodic Table", as in the arrangement set out in Table 18.2. This mini-Periodic Table consists of four rows and eight Groups, with Gd and Cm appearing twice. It reflects not only changes in ΔG°_{compl} by M^{3+} ions but also the tendency to show oxidation numbers other than +3. For instance, in Group 2 there is a tendency to form the +4 oxidation state and in Group 7 the +2, except for Am. As we know, the reluctance of Am to show the oxidation number +2 arises from repulsion between $4f$ and $5f$ orbitals and the relativistic effect. It is also interesting to notice that the sequence of colours characteristic of the M^{3+} lanthanide ions in the La → Gd series is repeated in the Lu → Gd series, i.e. that the f^7 configuration can also act as a symmetry centre. Thus, Pr^{3+} and Tm^{3+} ions are green, Nd^{3+} and Er^{3+} lilac; ions which have an empty (La^{3+}), half-filled (Gd^{3+}), or filled (Lu^{3+}) f subshell are colourless. Ce^{3+} and Yb^{3+}, which have one and 13 f electrons respectively, are also colourless.

Table 18.2 – Mini-Periodic Table for the lanthanide and actinide elements.

La	Ce	Pr	Nd	Pm	Sm	Eu	Gd
Gd	Tb	Dy	Ho	Er	Tm	Yb	Lu
Ac	Th	Pa	U	Np	Pu	Am	Cm
Cm	Bk	Cf	Es	Fm	Md	No	Lr

19

The transactinide elements

Quantum-chemical calculations show that with respect to their electron configuration transactinides, the elements from 104 to 112, form the fourth series of d block elements, having the $6d^n7s^2$ configuration. However, some transactinides may differ from their lighter congeners in Groups 4–12. Presumably the differences could arise from the exceptionally large direct relativistic effect, which stabilizes the $7s$ orbital, and from very large relativistic splitting of $6d$ orbitals into $6d_{3/2}$ and $6d_{5/2}$ orbitals. Comparison of Fig. 19.1 with Fig. 19.2 shows that for $6d$ elements, in contrast to $3d$, $4d$ and $5d$ elements, the energy and spatial order of valence $(n+1)s$ and nd orbitals is reversed. In all transactinides the $7s$ orbital has the highest value of the radius $\langle r_{nl} \rangle$, whereas its energy is below that of $6d_{3/2}$ in the elements from 104 to 106 and is below that of the $6d_{5/2}$ orbitals in elements from 108 to 112. This reversal can be ascribed to the same relativistic effects which reverse the ordering of energy and spatial extent of $(n+2)s$ and $(n+1)d_{3/2}$ orbitals in Lu and in some actinides, see Section 18.3. The reverse sequence of orbital energies and radii may have a significant effect on oxidation states of transactinides, because in the ionization process the first electrons are detached from the energetically outer but spatially inner $6d_{3/2}$ or $6d_{5/2}$ orbitals.

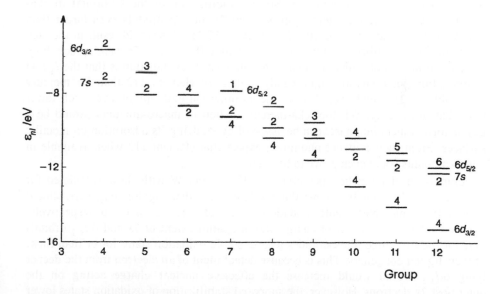

Fig. 19.1 Energies and occupancy of valence orbitals in transactinide atoms.

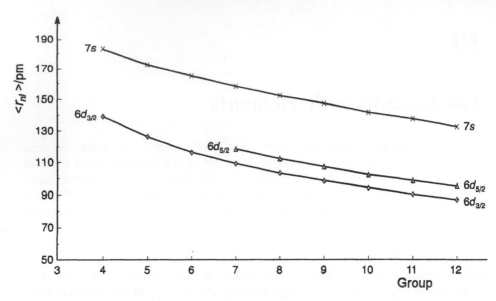

Fig. 19.2   Radii of valence orbitals in transactinide atoms.

Of course, transactinides cannot form condensed phases both because they are obtained in amounts of a few atoms and because they decay very rapidly, releasing large amounts of energy which would immediately vaporise the hypothetical solid. However, one can presume that in the hypothetical solid state all transactinides would show metallic properties and have high melting points, with the notable exception of element 112 (eka-mercury) which in this respect should be similar to mercury itself, its lighter congener. Strong stabilization of the $7s$ orbital in eka-mercury would make the energy gap between $7s$ and $7p$ orbitals even larger than that between $6s$ and $6p$ in mercury (see Section 17.1). This would result in the lack of overlap of the filled $7s$ and $6d_{5/2}$ bands and the empty $7p$ band precluding, thereby, formation of a metallic phase. An alternative explanation is that the I_1 and I_2 ionization potentials are significantly higher for element 112 than for mercury (see Table 19.2), which would make the atomization enthalpy of 112 much lower than that of mercury. Whether eka-mercury, when in macro-amounts, would be a gas or form a metallic phase, can be assessed by studying its adsorption on metallic surfaces. From such evidence one might expect that element 112 when available in macro amounts could even prove to be a gas.

Comparison of ionization potentials of Hf, Ta and W with those calculated for elements 104, 105 and 106 shows that the latter, like their lighter congeners, should exhibit the same most stable oxidation states, of +4, +5, and +6 respectively. However, the reverse order of energy and of spatial extent of $7s$ and $6d_{3/2}$ orbitals may for the first three transactinides lead to stable oxidation states lower than those of their lighter congeners. This is because detachment of an electron from the deeper lying $6d_{3/2}$ orbital would increase the effective nuclear charge acting on the outermost $7s$ electrons. However, the supposed stabilization of oxidation states lower than the Group number has not yet been confirmed experimentally. At the moment we know the chemical properties of element 104 (rutherfordium) relatively well from experiments, know some properties of element 105 (dubnium), and have a very limited knowledge of elements 106 (seaborgium), 107 (bohrium) and 108 (hassium). As far as chemical properties of rutherfordium are concerned we know for instance that it forms a volatile tetrachloride, which was the basis of its chemical identification. However, it has been found that the volatility of rutherfordium tetrachloride and tetrabromide is greater than that of the respective hafnium halides, probably because of the more covalent character of the $Rf-X$ bond, which may in turn be the result of relativistic stabilization of the $7s$ orbital. In general, rutherfordium accords well with the properties of its lighter congeners Hf and Zr. A good example is the ionic radius of Rf^{4+}, which according to calculations is very similar to that of Hf^{4+}. We also know from experiments that Rf(IV), like tetravalent Hf and Zr, easily forms anionic chloride complexes. On the other hand recent experiments suggest that rutherfordium shows hydrolytic properties more similar to those of Ti than of Hf and Zr, which may also be the result of relativistic stabilization of the $7s$ orbital. As far as chemical properties of element 105 are concerned we know from experiment that the volatility of $DbCl_5$ is similar to that of its lighter congeners and that the stability of its fluoride complexes in aqueous solution resembles that of tantalum. The very few experiments performed on seaborgium (element 106) indicate similar volatility of seaborgium, tungsten and molybdenum oxochlorides, and to formation of similar complexes with fluoride ions in aqueous solution. Recently performed gas chromatography experiments with bohrium (element 107) show that it forms a volatile oxochloride BhO_3Cl, just like manganese, technetium and rhenium, and behaves on the whole as a member of Group 7. The first-ever chemical studies of hassium (element 108), performed in 2001, have shown that it forms a gaseous oxide similar to that of osmium, albeit less volatile, confirming that element 108 is a member of Group 8.

Most chemical properties of elements 106, 107 and the still heavier transactinides can, at the moment, be deduced only from the basic properties of their atoms and are subject to considerable uncertainty. Because of the marked splitting of $6d$ orbitals into $6d_{3/2}$ and $6d_{5/2}$ orbitals (Fig. 19.1) heavy transactinides may differ from their $5d$ homologues in regard to stabilities of oxidation states. Since in elements 107 (bohrium) and 108 (hassium) the energetically highest lying $6d_{5/2}$ orbital is occupied either by one or by two electrons, the two elements could show oxidation states +1 or +2, which are uncommon for $5d$ elements in Group 7 and 8, respectively. The lower oxidation states could be relatively stable, because detaching the next electrons from the $7s$ orbital would be slightly more difficult due to lack of the deeper lying $d_{5/2}$ electron or electrons which screen the nuclear charge acting on $7s$ electrons. Because in element 111 (eka-Au) the $7s$ orbital is relativistically strongly stabilized there are, according to calculations, two electrons in its $7s$ orbital, in contrast to the $6s^1$ configuration of gold. The filled $7s$ subshell would make the electron affinity and electronegativity of eka-gold much smaller than those of gold and would preclude formation of anionic species similar to the Au^- anion. On the other hand strong stabilization of $7s$ electrons (high ionization potentials, see Table 19.1) in element 110 (eka-Pt) and in eka-Au would make the two elements even more noble than their lighter congeners, Table 19.1.

Table 19.1 – First ionization potentials (in eV).

Pt	Au	110	111
9.02	9.22	9.9^{a}	10.7^{a}

^{a} Calculated value.

As adumbrated earlier in this Chapter, it is highly probable that element 112 may differ significantly from its lighter congener mercury. Table 19.2 shows that the first and second ionization potentials of element 112 are significantly higher than those of mercury. Moreover, the estimated radius of the hypothetical 112^{2+} ion appears to be greater than the radius of Hg^{2+}, which means a lower hydration enthalpy. Both factors, higher ionization potentials and greater ionic radius, would make element 112 reluctant to form an $M^{2+}aq$ cation in the aqueous phase. Thus, eka-mercury could be not just a gas but moreover a noble gas reacting only with very electronegative elements such as fluorine and oxygen. That element 112 could be as noble as xenon is shown by the sum of its first four ionization potentials – almost equal for these two elements.

Table 19.2 – Ionization potentials and ionic radii of Hg and element 112.

	Hg	112
I_1 /eV	10.4	12.0^{a}
I_2 /eV	18.8	22.5^{a}
r_i (2+) / pm (CN 6)	102	115^{b}

^{a} Calculated; ^{b} estimated.

It follows from this review that in spite of some experimentally found and theoretically predicted differences, transactinides on the whole seem to resemble their lighter congeners in the 5d series rather closely.

20

The structure of the Periodic System

In previous Chapters we have discussed how and why fundamental properties of atoms such as orbital energies, orbital radii and ionization potentials determine chemical properties of elements. We intend in this Chapter to discuss firstly differences and similarities between the blocks of elements and then to recapitulate the main factors which affect chabnges of element properties down the Groups and across the rows of the Periodic Tale.

Grouping of elements into *s*, *p*, *d* and *f* blocks provides a very broad classification of elements within the Periodic Table. This grouping is shown on the Periodic Table printed at the beginning of this book (p. viii).* As frequently pointed out in previous Chapters, elements which have the same number of valence electrons but belong to different blocks, as e.g. elements from Groups 4 and 14, usually have very different properties. Therefore in the next paragraph we present a brief comparison between blocks of elements in regard to fundamental properties of their constituent elements.

The elements of the *s*, *d* and *f* blocks are electropositive, have low to moderate ionization potentials, and all form metallic phases, albeit differing enormously in atomization energies. In this context it should be noted that with respect to atomization energies *f* metals are intermediate between *s* and *d* metals. Also the number of oxidation states points to the intermediate character of *f* elements, because it is limited to one for *s* elements, is high for most *d* and *p* elements and is at most two for *f* elements (except for the actinides from Pa to Pu). On the other hand, *f* elements resemble *s* elements more than *d* elements in their reactions with non-oxidizing acids, water, and with respect to rate constants for ligand substitution in complexes. Such rates are, as a rule, high for *s* and *f* cations and very low for some *d* cations. The elements of *s* and *f* blocks form complexes (weak in the case of *s* elements) with multidentate ligands containing oxygen or nitrogen donor atoms, but are reluctant to form complexes with halide ions. In contrast, *p* block elements prefer halides to oxygen- or nitrogen-bearing ligands, whereas transition elements eagerly form complexes with both multidentate and halide ligands and in this respect are indeed transitional elements. In contrast to *d* elements, *s* and *p* elements have no subshells which could be split in ligand surroundings, whereas in lanthanides and actinides the *f* subshell has the required properties, but is very deeply buried. This has its consequences in the stability, spectroscopic and magnetic properties of complexes, because there is no ligand (crystal) field effect in complexes of *s* and *p* elements, whereas this effect is very strong in *d* ions and small or intermediate in complexes of *f* elements. In contrast to *s*, *d* and *f* elements, which

---

* Note that the assignments of some elements are arguable, for example He (see Sections 3.2 and 14.1), Cu, Ag and Au (see Section 15.1) and La, Ac (see Sections 9.4 and 18.1). In some forms of the Periodic Table Lu and Lr appear as *d* elements, replacing La and Ac in Group 3. Zn, Cd and Hg are not *d* elements, but are often discussed together with the *d* elements.

are all metals, the p elements are unique in that most of them form non-metallic solid phases or exist as molecular or atomic gases. Also most of them are highly electronegative and have high ionization potentials. On the other hand the p block elements resemble d rather than s or f elements in regard to the abundance of oxidation states. However, there is one important difference in that the number of oxidation states for d elements is the highest in the Groups around the middle of the block, whereas in the p block the elements on the right-hand side show the greatest variety of oxidation states. Another difference between p elements and the elements of the remaining blocks is the way in which the oxidation state changes, i.e. either by two or without this restriction, respectively. This survey shows that filling subshells with different quantum number l by electrons differentiates the elements, but that many properties change in different ways on going from block to block. This makes the classification into blocks less definite than into Groups. The p block best shows the conventional character of this type of grouping, because it comprises typical metals, semiconductors and non-metals. Also in contrast to s, d and f elements, the p block elements have both positive and negative stable oxidation states. Therefore, one can say that from the chemical point of view the p block is much more heterogeneous than the s, d and f blocks.

Groups are the principal units of classification and are composed of elements much more similar one to another than elements which constitute a block, except for the first element in the s and p Groups. Because f elements are very similar (excluding the early actinides), they are not grouped in accordance with the number of f electrons but are commonly discussed as a whole. However, as we know, properties of elements, in addition to the uniqueness of the first element, do change down the Groups and in this way form the primary and secondary structure of the Periodic System. By the primary structure or primary periodicity we mean the fundamental law of periodicity which says that chemically similar elements appear at intervals of the atomic number $\Delta Z = 2, 8, 8, 18, 18, 32, 32....$ As far as the last Period is concerned we know for sure that the elements from 104 to 108 resemble their lighter congeners in many respects. The experimentally observed periodicity in fundamental properties of atoms and in chemical properties of elements is produced by the dependence of orbital energies in many-electron atoms on both quantum numbers n and l. The dependence on l results from interelectronic repulsion and different radial probability density in the vicinity of the nucleus for the s, p, d and f electrons. In the primary structure of the Periodic System we can also include the general trends in changes of chemical properties down Groups and across Periods as e.g. the increasing metallic character of elements and increasing stability of lower oxidation states down the p block Groups or the increasing electronegativity and ionization potentials across the rows. However, we know from previous Chapters that changes within Groups and Periods are generally neither uniform nor gradual, and are quite frequently not even monotonic. We also know that down the Groups of p block elements some properties of atoms and elements even show oscillatory behaviour, commonly called secondary periodicity. This non-uniform, sometimes non-monotonic and oscillatory, behaviour of the properties of the elements constitutes what can be called the secondary structure of the Periodic System. The following four factors produce the secondary structure of the Periodic System:

– The unique properties of the first subshell with the quantum number $l = 0$, 1, 2 and 3, i.e. of the $1s$, $2p$, $3d$ and $4f$ subshells;
– The appearance of the first d series (the $3d$ series) which affects the properties of $4p$ elements
– The appearance of the first f series (the $4f$ series) which affects the properties of $5d$ and $6p$ elements.
– Relativistic effects which influence properties of the heaviest members in each Group.

With regard to the first factor we know that there is particularly strong repulsion between electrons in an $(n+1)l$ subshell and electrons in the lower lying filled nl subshell. As already explained in Section 3.2, the suggested reason for this repulsion is the same angular distribution of probability density, which makes it difficult for outer electrons to avoid the inner electrons with the same l quantum number. Since the first l subshell does not have a deeper lying subshell with the same angular distribution of probability density it does not experience this strong repulsion and radial expansion. This is the reason why the $2p$ subshell is unique with respect to its very small radius, almost equal to that of the $2s$ shell, and its strongly negative orbital energy. The $1s$ and to some extent also the $3d$ and $4f$ subshells show similar unique properties. On the other hand the second l subshell expands radially in order to diminish electrostatic repulsion, which makes it differ very much from the first and significantly increases its distance from the next lower lying subshell, which in the case of p elements is the s subshell. The distance between the np and ns subshells for $n \geq 3$ thus becomes more or less the same, see Fig. 3.3. As shown in previous Chapters, the lack of a deeper lying subshell with the same angular momentum quantum number makes the chemical properties of the first element in each Group very different from those of its heavier congeners. Examples are hydrogen, helium which in accordance with its electron configuration is the first element in Group 2, and all elements of the $2p$ series. Also $3d$ elements differ considerably from their homologues in the $4d$ and $5d$ series. Moreover, as shown in Chapters 7 and 8, the effect of the large differences between the radii of the $2p$ and $3p$ subshells in p block elements is transferred on to the s elements in the third and fourth Period. This results in relatively large differences between sodium and potassium and somewhat less pronounced differences between magnesium and calcium, with respect to both fundamental properties of atoms and chemical properties of elements (Sections 7.2 and 8.2).

The second, third, and fourth factors, i.e. the presence of filled $3d$ and $4f$ shells and relativistic effects, make changes down the Groups of p block elements, starting from the second element, non-uniform. As already mentioned in previous Chapters, incomplete shielding from the nuclear charge by the d and f electrons increases the effective nuclear charge acting on the outer s and p electrons in the elements of the $4p$ and $6p$ series, respectively. The effect is so strong for Group 13 – 15 elements that the s and $p_{1/2}$ orbital radii and orbital energies alternately decrease and increase in these Groups between the third and sixth rows. Inclusion of the unique properties of the first element in the Group extends the range of oscillatory behaviour into the second Period, see Figs. 5.3, 5.4, 9.3, 10.2 and 11.1. The lower values of radii and more negative energies of $6s$ and $6p_{1/2}$ than of $5s$ and $5p_{1/2}$ orbitals in Groups 13 – 15 elements result not only from the presence of the filled $4f$ shell but also from the direct relativistic effect which stabilizes the s and $p_{1/2}$ valence electrons. As shown in Chapters 11 and 12, the oscillating or non-uniform changes in fundamental properties of atoms down the p block Groups find their counterpart in non-uniform changes in some chemical properties of the corresponding elements as e.g. in the lower stability of As(V) than Sb(V) and in the instability of Br(VII). The presence of the filled $3d$ and $4f$ shells affects the outer shells not only in atoms but also in the M^{3+} and M^{4+} ions of Group 13 and 14 elements, respectively, see Figs. 9.1 and 10.1. It is seen in these Figures that the respective ionic radii increase less between the third and fourth and between the fifth and six than between the fourth and fifth Period. The non-uniform changes of the outermost shells in the ions are less distinct than in atoms, but are strong enough to divide the M^{3+} ions of Group 13 elements into the Al–Ga and In–Tl subgroups and the M^{4+} ions of Group 14 into the Si–Ge and Sn–Pb subgroups.

Due to the first and third factors the d block elements also show non-uniform changes of basic properties of atoms and chemical properties of elements down the Groups. For instance, due to the first factor, the differences between the $3d$ and $4d$ orbital energies and radii are distinctly greater than those between the $4d$ and $5d$ orbital energies and radii, Table 15.4 and Fig. 15.14. As discussed in Section 15.6 this has its chemical consequences, in e.g. greater stability of higher oxidation states in elements of the $4d$ and $5d$ than of the $3d$ series. On the other hand the similarity between the $4d$ and $5d$ shells results from the presence of the filled $4f$ shell which does not fully shield outer electrons from the nuclear charge. Since the $4f$ subshell is within the $5p$ subshell (see Fig. 18.1) its presence in $5d$ elements decreases not only the radius of the $5d$ but also that of the $5p$ subshell and makes its radius very similar to that of the $4p$ subshell. This, as we know, results in similar ionic radii for elements of the second and third transition series (see Table 15.5). As in the case of $3d$ and $4d$ elements, repulsion between $4f$ and $5f$ orbitals and relativistic destabilization of $5f$ orbitals makes light actinides differ considerably from lanthanides with respect to the stability of oxidation states higher than $+3$.

We know from previous Chapters that the p and d block elements differ enormously with respect to almost all their chemical properties. However, there is a striking resemblance between the p and d elements with regard to changes of some fundamental properties of atoms and related properties of elements down the Groups. Thus, both the $2p$ and $3d$ orbitals show distinctly smaller radii and more negative energies than the respective orbitals with higher n values (compare Fig. 15.15 with Fig. 3.3). The reason is the same, i.e. lack of more deeply lying orbitals with the same quantum number l. On the other hand similar properties of $4d$ and $5d$ orbitals correspond well with similar properties of $3p$ and $4p$ orbitals. The reason is again the same, i.e. the presence of a filled subshell, either $4f$ or $3d$ respectively. The end elements of the $5d$ series also show similarity to $6p$ elements with regard to the role played by relativistic effects.

Relativistic effects contribute to non-monotonic changes in fundamental properties of atoms and chemical properties of elements down all Groups. One example is the reversal in the general trend of decreasing ionization potential down the Groups for the Cs–Fr, Ba–Ra, In–Tl, Sn–Pb Ag–Au and Cd–Hg pairs. In Chapters 9 to 18 numerous examples were given of how relativistic effects make the heaviest element in a Group differ markedly from its lighter congeners, e.g. in respect to stabilities of oxidation states. However, it should be noted that in p block elements the direct relativistic effect which stabilizes the s electrons favours lower oxidation states, as e.g. thallium(I) and lead(II), while in d block elements the indirect effect (destabilization of the $d_{5/2}$ orbitals) makes the higher oxidation state more stable, as is the case with e.g. platinum and gold.

In rows of s, p, d and f block elements the general tendency is a decrease of orbital energies and shell radii, and an increase of ionization potentials and electron affinities. However, changes in these properties are not always monotonic. Examples are the lower first ionization potential of oxygen than nitrogen, and much lower electron affinity of nitrogen than carbon. The common reason is the increased interelectronic repulsion in the p^4 configuration which has one electron in excess of the half-filled p shell, i.e. three pairs of electrons with antiparallel spins. A similar effect i.e. lower stability of the configuration with one electron in excess of a half-filled d subshell makes the ionization potential I_3 significantly lower for Fe, Ru and Os than for Mn, Tc and Re, respectively. This has its chemical consequences in e.g. the much more positive, by about 0.7 V, reduction potential of the Mn^{3+}/Mn^{2+} couple than of the Fe^{3+}/Fe^{2+} couple. Among f electron elements the ease of detachment of the first electron in excess of the half-filled f subshell (low value of I_4, see Fig. 18.8) results in formation of Bk(IV) and less stable Tb(IV). Not only ionization potentials but also electron configurations of d and f elements change non-uniformly across the rows, but show transfers from s to d (from d to s in Tc) and from f to d, respectively. Among transition elements such transfers result in +1 oxidation state compounds of Cu, Ag and Au. Changes in configuration originate, as explained in Sections 15.6 and 16.1, in the energy gap between the relevant orbitals and in the exchange energy.

An important feature of the Periodic Table is that the identity period is not constant but, except for the first period, increases in pairs. On the other hand each of the s, p, d, and f blocks separately shows the same identity period equal to 2, 6, 10 and 14, respectively. In contemporary forms of the Periodic Table the s, p and d blocks are presented together, whereas for practical reasons (but inconsistently) the block of f elements is presented separately.

We may conclude that the organization of elements into Groups, rows, and blocks in the familiar Periodic Table developed from Mendeleev's original version now, as in the closing decades of the 19th century, provides the best means of systematising inorganic chemistry. In earlier times it indicated where and how to seek then-undiscovered elements such as gallium and germanium. Over the past 130 years or so it has provided a framework for the discussion and rationalization of the ever-increasing amount of chemical information and knowledge. In recent years higher actinides and transactinides have taken their places in its logical extension, and it may be confidently expected to accommodate the few yet heavier elements which remain to be discovered. Although in principle we are probably still far from the end of the process of building up elements by the continuing addition of protons, neutrons and electrons, in practice we have almost reached the point at which the life-times of the heaviest elements have become so short as to preclude the study of their chemistry in any meaningful sense – the Periodic System is approaching completion.

INDEX

Element entries are not comprehensive – only major and significant minor mentions in the text are cited here.

actinide contraction, 175-177
actinides, 1-4, 169-186
actinium, 17, 172, 174, 181, 191
activation volumes, 140
allotropy, 65
aluminium, 50-51, 54, 85-86, 89-95, 99, 183
amalgams, 168
americium, 169, 178-179, 182
ammonia, liquid, solutions, 73, 77, 102, 161, 169
antimony, 59, 65, 107-108, 110-113, 129
aqua-ions
  hydration numbers, 73, 82, 92, 95, 158, 166-167, 169, 182-183
  lanthanides and actinides, 183-184
  pK_a values, 92, 95, 112, 139, 167, 183
  substitution kinetics and mechanisms, 82, 93-95, 158, 167, 192
  transition metals, 137-140, 145-146, 157
argon, 127-128, 130
astatine, 121-122, 124
arsenic, 56, 59, 60, 65, 103, 107-108, 110-113
atomization enthalpies (energies), 60, 63

band gap, see energy gap
bands, energy or conduction, 61-66, 133, 156, 164, 170-172, 188
barium, 18, 77-84
berkelium, 176, 179, 182
beryllium, 3, 30, 77-82
bismuth, 4, 15, 42, 50, 65, 92, 104, 107-108, 110-112, 114, 122
bohrium, 189
boron, 4, 55-56, 60, 64, 69, 85-89, 99, 109
bromine, 32, 121-126

cadmium, 126, 163-168
caesium, 17, 18, 38, 42, 70-76, 126, 161
calcium, 77-84
californium, 2
carbon, 3, 56, 59-60, 82, 97-100
catenanes and knots, 159
catenation, 55-58, 101-102, 116-117

cerium, 18, 169, 179, 181-183
Chatt-Duncanson classification, 92-93
chlorine, 121-126
chromium, 16, 19, 119-120, 128, 146-148
clathrates, 130
clusters, 59-60, 101-103, 105, 111, 112, 114, 131, 153-154, 162
cobalt, 59, 72, 92, 110
complexes
  s block elements, 73-76, 82-84, 192
  p block elements, 90, 101-103, 106, 192
  d block elements, 106, 140-148, 152, 157-160, 165-167, 192
  f block elements, 182-186, 192
condensed phases, 60-61, 64-66
copper, 19, 117, 155-160
crown ethers, cryptands, 73-76, 83, 102
crystal structures of ionic compounds, 72-73, 81-82, 104, 118, 126, 139
curium, 2, 19, 169, 172-174, 178, 179, 186

diagonal relations, 72, 80, 85, 87, 111
dimers and dimerization, 52, 54, 61, 89, 90, 110, 157
disproportionation, 45-48
dubnium, 189

effective atomic number, 21-22
effective nuclear charge, 20-21, 176-177, 180
eighteen electron rule, 147
einsteinium, 3
electron affinity, 35-36
electronegativity, 37-38, 189, 193
electron configuration, 15-20, 94, 111, 131-151, 155-158, 163, 169-170, 178-189
energy (band) gaps, 64-66, 100-101, 117, 122, 161, 164
erbium, 176, 184, 186
europium, 170, 178, 181, 183

fermium, 3, 176
fluorine, 28, 48-49, 103, 112-114, 116, 121-126, 128-130, 157, 160, 182

francium, 17-18, 70-71

gadolinium, 19, 169, 178, 179, 181, 182, 184, 186

gallium, 25, 49-51, 54, 85-86, 89-94, 144

germanium, 4, 52, 56, 97-103, 105

gold, 19, 36-38, 42, 122, 155-162, 168

hafnium, 104-106, 152

Hard and Soft Acids and Bases (HSAB), 38-39, 92-93, 106, 157, 165-166

Hartree method, 9-14, 15, 19

hassium, 1, 189

helium, 2-3, 25, 127-128

holmium, 184

hybridization, 43, 49-50, 64-65, 109, 112, 116, 118-119, 123, 129, 168

hydrates, 81-82, 138-139, 169, 182

hydration enthalpy, 45-47, 92, 137, 145, 183

hydration numbers, *see* aqua-ions

hydrogen, 5-8, 12, 67-70, 72, 77, 87-90, 101

indium, 29, 34, 50-52, 54, 85-86, 89-94

inert pair effect on oxidation states, 45

iodine, 58, 121-126

ionization potential (energy), 34-35, 40-42, 47-53, 63-64, 94-98, 101, 104-106, 127, 133-135, 149-151, 156, 160, 163-167, 170, 178-180, 190

iron, 60, 110, 115-116, 122, 147-148

Irving-Williams stability sequence, 145

krypton, 11-12, 32, 127-130

lanthanide contraction, 175-177

lanthanides, 169-186

lanthanum, 16, 17, 69, 94-96, 169, 181, 182, 186, 191

lattice energy, 70, 72, 78, 116, 125, 166

lawrentium, 169, 174, 176, 191

lead, 26, 34, 42, 50, 52-53, 56, 59, 97-106, 122

lithium, 12, 25, 29, 30, 70-76

lone pair
    donation, 109-110, 116
    stereochemical effects, 54, 58, 91, 103, 112, 116, 118, 123, 125, 129

lutetium, 94, 174, 176-178, 180, 181, 183, 185, 186, 191

magnesium, 3, 28-29, 46-48, 77-84, 165-166

manganese, 28-29, 45, 92, 116, 122, 131, 142

mendelevium, 178

mercury, 34-35, 42, 45, 53-54, 56, 59, 117, 163-168

metal-metal bonds, 131, 144, 153-154, 162, 167

metallic phases, 4, 60-66, 105, 111, 122, 123, 132-134, 164, 170-173

mixed valence compounds, 90, 102, 114, 120, 181-183

molybdenum, 60, 119-120, 148, 152-154

multicentre bonds, 42, 69, 86-89, 112, 119, 123-125, 129

neodymium, 42, 169, 178, 182, 186, 189

neon, 127-128

neptunium, 2, 42, 169, 172-174, 179

nickel, 129, 144, 148

niobium, 19-20, 113-114

nitrogen, 53, 57-60, 72, 107-110, 121

nobelium, 2, 178

non-stoichiometric compounds, 69, 156

orbital
    energies, 11-13, 16-20, 33-34, 111, 124, 135, 156, 161, 164, 173-175

orbital
    filling, 10

osmium, 59, 116, 122, 132

oxidation numbers and states, 43-54, 131-154, 169-190
    ambiguous, 147-148

oxocations, 113, 169, 179, 183-184

oxygen, 3, 28, 57, 72, 80, 109, 115-121

palladium, 19, 42, 129

Pauli exclusion principle, 15, 149

Periodic System, 191-196
    primary and secondary structure, 193

phosphorus, 4, 43-44, 57, 59-60, 107-108, 110-113

platinum, 19, 42, 72, 115, 122, 128, 144, 156

plutonium, 2, 42, 169, 173, 176, 179, 181, 183, 184

polonium, 42, 115, 117-119

polyanions, 102, 112, 113, 117, 120, 124-125, 152

potassium, 16, 18, 25, 70–76, 90, 98, 133
praseodymium, 176–177, 179, 181–182, 186
probability density, 5–8, 11, 26–28, 41
promethium, 2, 42, 184
promotion energy, 45, 48–49, 111, 127–129
protoactinium, 19, 169–170, 176, 179–181, 186

radium, 77–79, 82
radius, 21–32
    atom, 21–26
    covalent, 26
    crystal, 30–31
    ionic, 28–32, 46–48, 50–53, 90–92, 94–95, 101, 105, 110, 135–138, 141–143, 152–153, 162, 173–177, 187–190
    metallic, 27, 101, 133, 156, 170–173
    orbital, 6–9, 12–13, 18–19, 40–41, 46, 48, 50, 111, 142–143, 149–153, 156, 161, 171–177, 180–181, 187–188
    shell and subshell, 21, 47–48, 53–54, 135–137, 142, 148, 150–153, 175–177, 181, 193–195
    van der Waals, 31–32
radon, 2, 127–129
reduction (redox) potentials, 73, 78, 117, 137, 139, 146–147, 167, 183–184
relativistic effects, 34–37, 40–42, 70, 92, 101–102, 106, 119, 156, 160–162, 164, 167, 174–175, 186, 195
repulsion, orbital, 24–25, 128, 134–135, 180, 186, 193, 195
rhenium, 59, 122, 131
rubidium, 17, 70–76
ruthenium, 107
rutherfordium, 189

samarium, 42, 178, 181–183
scandium, 17–20, 94–96
screening (shielding), 19–21, 33–34, 37, 90, 128, 149, 157, 166, 172, 175, 177, 180, 194
seaborgium, 189
secondary periodicity, 52, 90, 101, 111, 124, 193–194
selenium, 115–119
semiconductors, 3, 90, 100, 122
separation, 75–76, 177, 184–186

shell and subshell filling, 15–20, 33–36
shielding, see screening
silicon, 3, 24, 52–53, 56, 97–103
silver, 4, 122, 126, 129, 155–160
Slater's rules, 21–22
sodium, 12, 18, 25, 70–76, 126
solubilities in water, 70, 72, 79–80, 89, 91
stability constants, 74, 76, 92–93, 145, 158, 165–167, 184–186
strontium, 29, 77–84
structures of binary compounds, 117–118, 126, 138, 181–182
sulphur, 4, 43–44, 49, 53–54, 57–58, 60, 115–120, 181

tantalum, 113–114
technetium, 2, 148–151, 154
tellurium, 4, 115–119
template reactions, 83–84, 158–159
terbium, 122, 176, 179, 181–182, 186
thallium, 26, 34, 42, 47–48, 50–51, 54, 85–86, 89–94, 106, 126
thorium, 2, 3, 181–183
three-centre bonds, see multi-centre bonds
thulium, 186
tin, 34, 52–53, 56, 59, 66, 97–103, 105–106, 110
titanium, 105
transactinides, 187–190
transition elements, 18–19, 131–154
tungsten, 59, 119–120, 133, 148, 152–154

uranium, 3, 42, 104, 169, 173, 179, 181–184, 186

valence, valence state, 43–45
valence shell electron pair repulsion theory (VSEPRT), 118, 123, 125, 129
vanadium, 19–20, 106, 113–114
van der Waals
    forces, interactions, 56, 59, 65, 111
    radii, see radii, van der Waals

xenon, 127–130, 151

ytterbium, 170, 178, 181
yttrium, 17, 94–96, 186

zinc, 29, 117, 163–168
zirconium, 29, 69, 104–106, 152

Printed and bound by CPI Group (UK) Ltd, Croydon, CR0 4YY

03/10/2024

01040435-0019